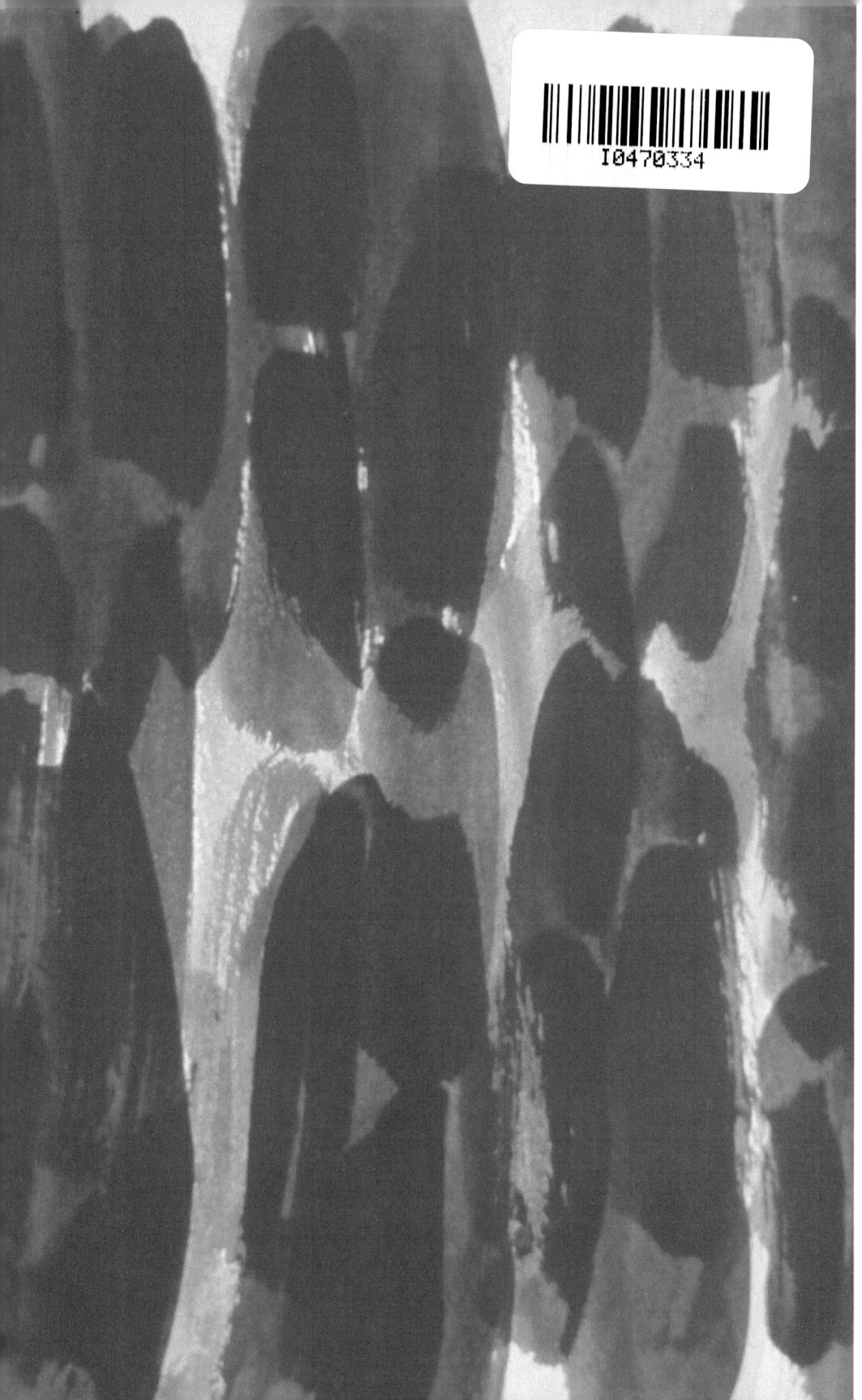

COLORS
Passages through Art, Asia and Nature

2nd Edition 2012

Text by Sarah Sutro
ISBN: 9781456373337
Library of Congress Catalog Card Number: 2010918528
Published by Blue Asia Press, N. Adams, MA (USA)

www.sarahsutro.com
E-mail: *artssutro@yahoo.com*
Text © Sarah Sutro
Book Design by Cynthia Friedman
Originally titled and distributed as:
Iron and Molasses: An American Artist Reflects on Natural Color

COVER AND TITLE PAGE IMAGE:
Marks IV by Sarah Sutro
cutch and indigo on paper, 18x14", 2001

LAST PAGE AND BACK COVER IMAGE:
Stems #1 by Sarah Sutro
natural color on paper , 12x9", 2004

Several chapters from this book were adapted as "Indigo, Asia's Blue Gold,"
in *Sawaddi Magazine,* 4th Quarter 2009; and as "Stories of Indigo," in
Lesley University's online *Journal of Pedagogy, Pluralism and Practice,* #10, 2006.
www.lesley.edu/journals/JPPP/10/sutro.html

COLORS

Passages
through
Art, Asia
and Nature

By
Sarah Sutro

Table of Contents

Acknowledgements

Thanks to *Aranya* and its staff, especially Ruby Guznavi, Moni Khan, Chamle, and Mr. Hussein, for their teaching and example. To Bangladesh, for giving me a glimpse into an older, more traditional culture. To Michael Bedford and Will for the gift of time, critical appreciation and for their continuing support and love. To Ellen London for her help in organizing the chapters. To my writing group in Dhaka, Marjorie Bucknor, Razia Sultana Khan and Bonnie Auslander, for their encouragement and response. To the Asia Study Group, especially Virginia Reekie, who took me to farflung places in Bangladesh. To Dr. Richard Weiner, Dr. Eleanor Pitts, Dr. Beverley Bowker, Heather Flaherty, the support group of Winchester Hospital, and friends and family who sustained me through medical treatment. To my parents and brother, who encouraged my relationship to nature. To my old friends in Ithaca and Boston who helped to make me who I am. To Joy and Dale Truman, who patiently took me to Caprilands Herb Farm again and again. To Ada Bortaluzzi, for getting me started. To Alka Mathur, Rokeya Sultana and Arham Chowdhury, who painted in natural colors with me. To the library at A.I.S.D. for providing me with my first written information about herbs and history, and the Duxbury group, who by their example prompted me to take my writing seriously. To Gaynor Blandford, who kept telling me I should write a book about Asia; Judith Cohen, for reading the manuscript and offering advice and support; Paula Bonnell, for her belief in my work; Tina Feingold, for her vision and good sense; and to Frances Hamilton, for her loving encouragement. To the Pollock Krasner Foundation that supported my work; to Harriet Barlow and Blue Mountain Center for their ongoing gifts of time, friendship and support; and to my colleagues in the Emerson College and Lesley University IRO faculty, who helped me develop back into a writer again.

Preface

Almost the first thing a visitor notices about South Asia is the color—
the butterfly swirl of a cloud of women in saris, each one more vivid
than the next; the very nearly garish (to a western eye) decoration of
the smallest Hindu temple or shrine; the bright dot in the middle of a
woman's forehead. In Bangladesh, where this book takes place, there are
more Muslims than Hindus, and colors are perhaps a little more muted.
But look at the trucks—they're painted every screaming shade you can
imagine. Look at the back of the snorting two-stroke taxis and the bicycle
rickshaws: each one has a bright painting of astronauts or animals. Color
almost assaults you in Dhaka, just like the humidity and traffic and smell.
Sometimes it looks like one of those early color tvs, where you could turn
the tint knob all the way to the right and the hues would just explode.

Now, "local color" is one of those phrases that journalists toss around—if
you're covering the president's visit to Omaha, getting local color means
you stop in the diner on some downtown corner to get a few lively
quotes from the people hunched over their coffee. But Sarah Sutro
gives the phrase much richer meaning. This book is quite literally about
local color—about where color comes from, or came from before it was
synthesized and globalized in the same manner as food and music and
pretty much everything else. When that happened, of course, much of the
meaning drained out of color, just as it drains out of everything else.

But perhaps that meaning didn't disappear for good. Sutro tells the
fascinating story of local Bengalis who are rediscovering the roots—
literally roots, and blossoms, and petals—of their colors. Who are growing
indigo again, for instance, this time not in vast and dire plantations, but as
small attempts at sustainable agriculture. Who are figuring out once more
the old recipes for boiling and setting dye. Who are fashioning a new
textile and decorative art from the shades of their past.

It's a story being told in a thousand other places around the world with
a thousand different commodities—the Slow Food movement in Europe

rediscovering the local sources of great taste, for instance. Always it's a movement against the gospel preached by the globalizers—the gospel of efficiency over all, where "lettuce" comes from California and movies come from Hollywood (or Bollywood) and "yellow" comes from the spigot in some pigment refinery. "Many times I felt impatient, weighing the tiny heads of flowers and dried bark, waiting for a mixture to boil, or cloth that was hanging in the sun to dry," Sutro writes. "Working with natural color can be a slow process, analogous to the process of healing." And indeed her book is interwoven with wonderful meditations on many ways of learning to live a little more slowly and a little more deeply, to listen a little more carefully to one's own heart, one's own neighbors.

I wish I could say that this approach to the world will carry the day. In truth it's hard to know. The momentum of the globalizers—the Walmartization of the planet—is incredibly strong. It's poignant to remember that within a mile or two where Sutro was patiently dying cloth, most of the word's cheap t-shirts are produced by young women working in Dickensian conditions. But it's too soon to call it a loss, either. Accounts like this make it abundantly clear what we sacrifice on the altar of efficiency; they add up to an indictment of our particular obsessions, and more importantly to a vision of something much richer.

Once you've read this book, your eye will take in more than it did before. And your heart perhaps as well.

Bill McKibben,
Ripton, Vermont, January 2004

Foreword

In this book about color, I trace my own experiences with color, painting and dye, from my professional experiences as an artist to the more elemental source material of memories, places, experiences, and especially, Nature.

I express my own developing awareness of nature, color, color dyeing, color association through myriad strands of a web, into the past, to other countries and back into history.

Bangladesh, where I have had the privilege to live for two years, serves as a catalyst for the stories I have to tell. Almost pre-industrial in its feel, materials are used unprocessed, and everything is recycled. In a shop not far from my home, I encountered a renewal of natural dyes, took a workshop, used the dye pastes to paint and draw, and absorbed the lessons of herbs first hand. I describe the workshop I took, for those interested in the actual process of extraction and dyeing.

In watching the dyeing of indigo products I learned about the history of that dye; reflected on my own encounter with the ghost of indigo production on a sea island in Georgia, and learned basic principles of making color fast. Indigo is one of the oldest known dyes. Indigotin, found in indigo and woad, when fermented, makes up the blue we know as indigo blue. Crushing the leaves with an alkali (base), plant ashes, wine or limewater, creates a product which can then be used to dye cotton or silk.

I debated, in organizing these stories, whether to form a grid, a chronological sequence or leave them as a tangle of garden. In a way they are a living history of colors. On a wheel, each one leads the reader in to the center, which is nature, or out to the periphery, which is how we experience and use it. They tell the story of my enlightening encounters with the powers of color. Sprinkled throughout the narrative are recipes and lists that may be interesting to the student of nature, color and paint.

Each locale, globally, has its own range of plants, herbs and flowers that afford natural color. In my stay in South Asia I have been influenced by what I have seen and encountered. But it takes longer than two years to really know a place's colors, plants, artists.

What I do know very well are very basic American herbs and flowers: basil, mint, morning glory. The colors and internal realities of these plants nourished me in childhood and adulthood into an awareness of my connection to all things natural. When my family left the outer suburbs to drive to the country, to the old farmhouse, where we spent our vacations, we moved into another world, the world of plants. These ordinary American plants begin the stories, and they conclude them.

Sarah Sutro,
Dhaka, Bangladesh, 2001

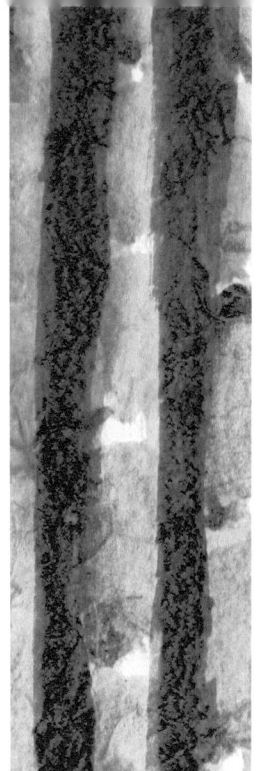

PART 1

Introduction: Of Dyes, Nature, Flowers, and Phytochemicals

I am beginning chemotherapy and I dream my surgeon is telling me to eat extracts of fruits and vegetables. He has not an alternative bone in his body, insomuch as I can see, yet he has a way of appearing in my dreams with wise sayings, and he has a knack of saying the right thing in real life, just when I need it. For instance he appeared in my hospital room the second night I was there, sick from too many drugs and struggling to pull myself together after surgery and offered to bring me an easel or paints, paper, pencils, whatever I needed. Where did he come from, I wondered? But on the subject of natural cures I also dreamed that there was a yellow, mustard-like flower, which may be marigold (also related to the color of turmeric), which is a cure for cancer. Can one take these messages seriously?

I lean over the pot of bubbling spaghetti sauce and think red, the color red. Red is supposed —or tomatoes at least—to help with cancer. I try to notice how I feel as I drink tomato juice and eat spaghetti. There was a time when my paintings were all red—and to me this indicated a sense of life energy that was new to me. Since then I have spoken with many friends who also went through their "red" phase, often coming out of a shy, or withdrawn time in themselves or their work.

What is it about these colors? I read in new medical literature that colored foods have phytochemicals—the greens of broccoli, the oranges of squash, the yellows of fruit, the purples of grapes. To eat colors is to eat health. To view colors is to be healed. I eat deliberately, choosing my palette with care.

In our current front yard north of Boston a yellow flower threatens to take over. At first we marvel at how bright and beautiful it is, a sharp, cool bright yellow, just on the way to grey-green. When it dries the flowers turn into small seed pods that rattle. We note this too with detachment. And then suddenly they are all dry and being blown all over the place and before we know it small clumps of this yellow flower are cropping up everywhere, elbowing out more delicate varieties, threatening to take over the whole garden, a reminder of cancer. We go out aggressively after weeks of contemplative studying, and start pulling out the yellow flower in clumps, but of course we are also distributing the seeds. After the yellow flower is in check, another flower starts to achieve ascendancy, and I notice the tall, dark purple flowers of columbine appearing too often as I scan the garden. So this is all about balance. The color, the recycling, the observation, the healing, the composition of a life.

From where we used to live in Somerville we would walk down to the end of the street, take a right through a small lane that ran behind a factory, then a left on Chandler. Right there was a small Indian food shop, which also rented out Indian videos. Michael and I would have planned an Asian meal on a Friday night, and we would buy saffron and turmeric and fresh *samosas* and walk back with our small plastic sacks to our two family frame house that was in the outskirts of Boston, just beyond Cambridge. There we would pour wine in the upstairs back kitchen, saute onions and garlic and various spices that formed a thick colored paste in the pan, and add the chicken. Later the yogurt and saffron, dissolved in hot water, turned the yogurt a deep and sweet reddish-yellow. Cover the pot, sit at the wooden table and wait until the smells started to come out of the deep iron pot with the heavy lid. The stove was an old one with a heater box that a friend had salvaged for us in her rental backyard. It served as the only heat for the kitchen. The first person down in the morning turned the heater on and it generously began pouring out heat from the side, almost immediately. Later we opened up the back of the kitchen, put on a small deck and sliding glass doors. Then the sun warmed the kitchen, too, except on frosty late winter mornings when nothing worked but the stove and hot tea and a wool scarf around the neck.

Much of the rest of the kitchen was appointed in the same way as the stove. The cabinet over the sink was scrounged from a trash heap on the way home from a drive out route 2. It had glass doors and looked just the right width. We gently fitted it into our trunk, leaving the trunk hood open, and after measuring it and making sure it would fit into the space, took it down to the basement to paint. It turned out to fit the kitchen perfectly, and after we painted the walls a twilight blue, the room had a mellow, assembled quality like a much used bulletin board, or a back mudroom that is both functional and well worn, loved, without pretension.

The living room we painted a Burmese Bush orange, after much discussion and many chips taped to various pieces of woodwork around the room. We found a large bright green rug in a yard sale, and the place felt complete. A fish basket for a lampshade in the hall, a homemade heating pipe invented by Michael leading to the attic bedrooms, a small antique table that was a wedding present as a spot for the telephone.

Many of our decisions seem to be made because of color and now my surgeon, in a dream, is telling me to drink color.

And then, we had a chance to live in Asia. Michael was offered a job at the Peace Corps. I was ready to stop teaching and have a chance to paint and write full time, and fully recover from my recent experience with cancer. Our son would go to an international school. When I discovered the natural dye store, *Aranya,* (which means forest in *Bangla*), close to where we were living, with its charts of herbs and flowers and the colors which resulted from boiling them and using them as dyes, I felt I was home. Here was indigo (its name seemed to come from another century), henna, marigold, frangipani, pomegranate, dahlia, onion, eucalyptus and many other plants whose names I could not pronounce labeled on a grid on the wall, soft and pure colors, like a late Mondrian of Asia, leaning towards the yellows (lemon, cadmium, neutrals, greens) with the one blue, indigo, thrown in at the upper right. Each square had a kind of integrity, being not exactly a color you would see on a color wheel or a paint chart, each of them rounding out into something else, a natural host of colors, as none, like the cadmiums and cobalts of oil paint, were made from metallic ore. I imagined that each of these might have been a note, and if so the whole chart might have sung itself like a giant chord, having a range from deep (the dark brown and dark blue), to light (the nearly brilliant yellows). At first it was the charts of dyes and the fabric which drew me – the clothing itself was of such a profusion that it was almost hard to know how to begin to look, the Eastern *shalwar kameezes,* the long silk lengths of *saris,* men's *kaftas,* scarves of every length and design, vests, blouses, pillow covers, bedspreads, handmade paper—all repeating and slightly changing the combinations of colors you could see on the charts, but in endless variety and pattern, like a Bach fugue that goes on and on, only slightly changing the configuration of its notes.

One Saturday I signed up for a workshop with the Asia Study Group at *Aranya* to learn more about indigo dying. In the workroom a long sixteen-foot table was set up with white cloth pinned to it. At one end a man in white was dipping a wooden block into an indigo mixture, then carefully pressing the block onto the cloth. Gradually, slowly, he built a pattern across the hem of the cloth, then began again to create a second row of pattern. In the back courtyard were two pots of dye, boiling. One of madder, a dark rusty red into which another man dipped a *batiked* cloth, pushing it in again and again with a long stick. Further back in the courtyard was a vat of indigo, the most costly of dyes, into which a long cloth was being pushed, coming up a deep, satisfying teal green, only to turn to brilliant chalk blue as the cloth hung on the line in the air and sun.

Nature seemed to be in the ascendancy in all this process. Nature, who created color. Nature, who presented a fan of plants and flowers for us to discover the uses of, the phytochemicals my surgeon was telling me about in my dream.

I lingered over the names of the plants I didn't know on the chart and their colors: *manjit,* a warm brick red; *sheuli,* a light gauzy yellow; *babla,* a pale pink orange; *shialnutra,* a greenish taupe; *pawa,* a pale rose; *holud & hartaki,* a deep dark purple-brown; *coran,* a coral red; *catechu,* a dark red; *hartahi,* gold; *shilkorai,* a brownish-orange; *latkan,* a bright orange; *sundari,* a purplish brown; *catechi-4,* a warm brown; *hartalki-4,* a shade of black. I wondered about all these plants which we of the West don't know, and what other uses their natures suited them for. Who is to say there aren't tonics among these colors that are known in the East, which we have yet to hear of? Naming them seems to give them power.

These plants and colors are the inspiration for stories. My thanks to *Aranya,* Bangladesh and the Asia Study Group for the chance and the reason behind telling them.

Iron and Molasses

Strength and sweetness. Hard, and soft. Dark and dark. In a corner of the outdoor shed where the *batiking* took place at *Aranya* was a small iron pot. In it people threw bits of discarded iron. They also poured in thick, dark molasses. After awhile the iron came off into the liquid, the molasses color mixed in, and a deep, rich grey-black dye was created. Just the intense heat and sunlight, which could be pretty high in Dhaka, forming the dye. A self-creating dye, like the ink of a squid or the dark silt-colored water of an old pond. Swimming in the Swift River as a child and being afraid of the edges where the fine grained silt collected, what lingered there?

The darker aspects of life so much harder to accept and make part of the larger pattern. How did we come out of childhood, some of us, not expecting there to be dark patches? The several years I seemed to court black, wearing it, writing about it, seeing life like an eclipsed moon or sun, in shadow, so that bright color was not even a participant in my days. Color came as a kind of healing, later, discovering in particular the color red, its aliveness and purpose and direction.

Yet the dark times had their sweetness, too. The long journey to England to find myself, the explorations, the people met and befriended. In England I seemed to meet mad poets. Perhaps it's because I walked everywhere, especially through parks. There they would be, on a park bench or at a jumble sale, as if I had on a sign that read "seek me out." Geoffrey in Hyde Park, going on and on about himself. Strutting back and forth over the grass, carrying on about his parents and his education. The absurdity of this full grown man ranting about being a child, yet eloquently reciting to me his poems about death and maturing. We compromised on a cup of coffee in a corner shop, sharing an hour, a day.

I remember the sweetness of the walk to Notting Hill Gate from Holland Park, where I rented a room—a walk through the park, looking for peacocks, and then turning down into the mews and small cobblestone streets until I got to the art school I was attending. There I worked in a large lower studio with skylights where about twelve other senior students and myself had partitioned (but not walled off) spaces, each with our easel and table for paints.

There I met Jon, and Moische, and Amelia and Catherine and shared times and cups of tea with them, especially in winter when we stayed warm by any means possible. Sometimes Jon would invite me to have a coffee outside, so we would lay down our brushes and go off into the cold, sitting down in a local coffee bar with teal and yellow tablecloths. It was cold, and I was very careful with money because I had only come with $2500, to last as long as possible, earned from as many jobs as I could hold down in the five months before I left college, after graduation.

Iron for the marriage of the couple I lived with in London, conceptual artists in a bad way with each other, their marriage pulling apart like old fabric. John older and more established, Barbara younger and finding herself after many years of marriage. A dance around the wooden table in the garden level kitchen, John placing the block of cheese with great authority in the center of the table as he proclaimed his latest philosophical statement, Barbara cutting a slice off the cheese as she countered his assertion. Back and forth, the cheese diminishing in front of them and the argument growing more heated. Barbara, with heavy black makeup, going off to Portobello Road to buy daffodils, bread and milk, something for supper. John, his hair wild, insisting that they do everything together. Molasses for the wild and silly times I had with my art school friends, in London. The rides to Shepherds Bush to frequent a certain pub, the outings outside of London, to walk, to see shows.

Ruby is the manager of *Aranya*. She had the vision of starting up the long dormant vegetable dye business in *Bengal* (the former name of Bangladesh and west India), and she created this now thriving shop. She is energetic, about my age, positive. With an imperious manner, she has the reputation of being a tyrant, although I rarely see this side of her. There is no doubt about the passion she brings to her work, she will do anything to make this small business make its mark. She says dealing with the international markets is too difficult—too much paperwork for a small place like hers - too many orders at once for a small shop—and yet she continues to promote this corner of the Bangladesh economy, to cheerfully restore long lost arts to a contemporary place. The shop is located on a busy road just past the bridge going into *Banani*. There is a small courtyard and two levels of workshops and rooms exhibiting the cloth and clothes made with vegetable dye. On the ground floor is a kind of showroom of just fabric; behind it are the workrooms with big tables, doors opening out onto backyard space where the dyeing takes place in the sun, and in the back, her office. It is here we all sat and had tea and chatted about the dyeing process, and Bangladesh and what we had seen. I recognized many faces from the expatriate community, many people, like me, who are drawn to the Asia Study Group to learn about the life here, the texture, the way things are done. We are all fascinated by being here, in the present, observing a society making its way despite severe hardship. Trying to right itself after an unnaturally difficult and hard birth in the uprising in 1971. East Pakistan, far removed from its difficult twin in the east, declared itself independent and reverted to its own, older language, *Bangla*. The Language Movement, a revolt against the Pakistani Urdu language and an arbitrary and unnatural land division during the Partition of India, carried Bengalis on a wave of passionate enthusiasm for their way of life. India helped defeat West Pakistan, and establish Bangladesh from what was East Pakistan. But there is darkness underneath daily life—people killed, industries ruined by the war. Yet the enthusiasm for life, the love of language and independence is palpable among the Bangladeshis. Iron and molasses.

Michael is out of town, Will is supposedly going bowling with Rosario, and I have a rare opportunity to soak up Bangladeshi history and culture. We learn that until 1856 only natural dyes were used for cloth both in India and Bengal. After that year when synthetic dyes were created, the long used natural dye activity in both countries and around the world, diminished. In 1861 alizarin took the place of madder root, the original source of all reds. And indigo was developed synthetically in the early 1900's, which quickly shut down the natural sources of indigo, which were much more time consuming to prepare. India, Bengal and the Americas were the primary early creators and exporters of Indigo. So Ruby's effort to restore Bangladesh's dyeing activity is significant. The plants used for dye are all growing wild in the countryside; the potential is there; without concern and focus the knowledge of these colors could completely drop out of circulation, another loss from the escalation of industrialization and synthesizing of culture.

Some days I go over to the store just to look at the colors and clothes—their artistry gives me a lift—and to look at all the new variations that are constantly being created and displayed. At first the range is overwhelming and I can't distinguish one pattern from another, but eventually I see the way one color works to another, and I begin to see the eye of an artist at work in the productions I view. This is a phenomenon I recognize from my own art—a certain need to slow the eye down in order to recognize and appreciate relationship and difference between colors and patterns. It is a question of harmony, I think, how all these colors work together and whether we are able to see them individually, or collectively as a mass or riot or confusion.

One day I call up and take Ruby up on something she has offered: to give me some dyes that I can play around with on paper. I bring two small plastic containers with me, which she fills with the remainder of some indigo, used in the store the day before, and black—from iron and molasses. I take these precious two containers back to my makeshift studio where I quickly cover a white page of paper with marks reminiscent of my drawings. I find the paper a bit too smooth—it is a paper I use for my ink drawings. I'm used to suspending the marks of ink on top of the surface of the paper—but now I find that I want the dye to act differently, to be absorbed, and so I find myself drifting away to work on something else, and later I cover up the dyes with their plastic covers. A couple of weeks later I go out to find the handmade paper store called The Source. It hosts a Mennonite project for helping Bangladeshis in the countryside make and sell handmade paper and other objects from water hyacinth paper. As I discover after poking around the store for a while, they also make papers from banana and jute. Before long I have selected many sheets of varying textures and gradations of natural color and sizes—and head back to my studio to try out the dyes again.

The dyes work much better on this paper. The solutions have a slightly syrupy consistency, which makes the color pull away from the edge of the mark in an interesting way. The surface of the papers is rough and receptive, and I find if I add water to the brush a soft hazy mark will sink into the paper, but if I add no water it will lightly rest on top of it, and the combination gives a sense of space to the work. The resulting drawings look like water. I keep going back to the dyes and the paper—just exploring, I keep telling myself, whenever I think of it, and one day I discover the indigo pot covered with mold, and throw the rest of it away.

Now I am just left with black. It sits on my windowsill and I draw off it whenever I think of experimenting, and my stock of paper dwindles. It isn't exactly my usual work I am doing—it has a different edge. I am seeing a new level in darkness, a possibility I hadn't explored. Similar to the time I worked only in subdued colors, to understand color. Through limitation came a deeper awareness. And it seems to keep my creative fire alight to be trying something new.

Tea

Tea is one of those substances which appears to be one thing but which is really much more.

My grandmother's wedding dress was formal, with many layers of ruffles on a high bodice, huge puffed sleeves and a long, full skirt. She was married to my grandfather in the Portsmouth Navy Yard, much to her distress. She and her ministerial candidate beau would have preferred a quiet church wedding, but her father, an admiral in the Navy, vowed that the oldest of his five daughters and the first to be married would have a full Navy wedding. So my grandmother and grandfather, Esther and the shy Frederick, were married on the Navy base with a big brass Navy band, a full contingent of midshipmen saluting and a gauntlet of swords to walk through as they departed for their quiet existence at various semi-rural Massachusetts towns. After the wedding my grandmother's wedding dress, sophisticated for her day, was put in storage for future generations.

And, generations did pull out the dress and use it. The wedding pictures of both my mother and my aunt show them in the full-sleeved, ruffled-front dress. Satin and sheer organza, tucks taken in the waist and the bosom. The dress had two parts, a skirt and the bodice itself. The gown was exquisitely sewn and designed, which is partly why it lasted so long.

When it came time for the next generation to be married, the memory of the dress hovered over all the preparations. My sister was to be married at twenty, to her college boyfriend, before he was sent to Fort Dix. The Vietnam War was on and everyone was uneasy that he would be sent to Asia. My sister lived warily at my parents' house, preparing for the wedding. It would be a modest one - at the old Unitarian church at the center of the small town where they lived. Friends would bring refreshments for after the service. No liquor served, just punch. Out came the dress—there was almost no discussion.

My mother was a woman of not only thrift, but tradition. You could see her take on the mending and appraisal of the wedding dress with her usual forthright, determined style. It was brought out of storage in the packed and tangled attic that housed my bedroom at the end under the eaves. I had insisted on making the room, carving it out of the cluttered space with the help of my father when I was fourteen and feuding with my sister. We had shared a room our whole lives, but I suddenly insisted on my own space, needing to order it in my own way and feel independent, and she was delighted as well to pick her own new green-sprigged wallpaper and turn out the light when she wanted to, not to mention turning up the AM radio. Her desk was weighed down by calculus and physics books and mine with art and poetry, and we were each happy.

The biggest problem with the dress was its color. Originally white, the 50 years or so of its Rip van Winkle existence had left it a different color in the creases than on the folds. It had an aged, tired look, although the structure and the sewing remained strong. My mother thought about this for quite awhile. You could see it was on her mind as she went about her tasks in the kitchen, placing a pot lid on a double boiler with unusual decisiveness and whisking around the kitchen looking for a sponge. At last she came up with an idea: she would dye the wedding dress with tea, letting it soak slowly in a medium, medicine-brown brew until the whole article took on the same off-white cast. She did samples, first, placing squares of similar fabric in a saucepan of diluted tea, timing the dyeing, holding the finished piece up to the light to see how she liked the color. Eventually she settled on a very soft, light coffee color, the subtlest of tans, and with great care boiled a huge pot of water on the stove in the largest container she could find: the lobster pot, dipping several tea bags tied together at the ends into the water to achieve the desired degree of color, and then gently, gently lowering first one piece of the dress, then the other, into the cooling brew. She poked and stirred the tea water and fabric with a wooden spoon, and after the allotted time brought it out, dripping, to hang on the clothesline on top of a series of towels. Success. The dress came out a perfect pale coffee cream, like the coffee cream sponge dessert that she so liked to make, with the mold shaped like half an Easter egg, when guests were coming.

For the wedding my sister wore the long train that came with the dress (this had to be dyed separately) held on by a circlet of flowers, and held a small bouquet of roses put together by a woman in the church who did floral designing for a living. She was the first female cousin to be married. She was setting precedent.

One of my cousins also wore the dress, although I was out of the country when she got married, but I had the idea forming in the back of my mind that (if I ever married) I would never put it on. My sister had been of the generation of young women who married or became engaged in college, high school or just after. In contrast, most of my friends hadn't married right away, some never, and our twenties expanded out into an entropy of experimentation and short chapters of life, as the war and political unrest and changing societal mores raged all around us.

I was married at 33, on a beach on an island off Massachusetts. I didn't know till the day of the marriage what I would be wearing. We had arrived by ferry, my political activist/writer boyfriend and I, with our bikes and backpacks on board, having entertained my family at a dinner at our apartment the night before. My mother brought lilacs from her yard which we placed all over the apartment, in urns and coffee cans, so that the whole place smelled like May, and I served chicken with fresh rosemary, served on mismatched plates. Both of us loved islands, so we chose the closest one to where we lived, to be married on, Martha's Vineyard. We pushed our bikes off the boat, and started the six mile ride by the coast which would lead us to a hotel, and our appointment with the woman justice of the peace we had chosen over the preacher with the top hat and tails. The next day, realizing we needed something special to commemorate the time, we went browsing among the few stores that were open in the pre-season cool May morning. We sampled thrift stores, we looked in craft stores, and even bought a bouquet of flowers from someone selling it outside. In one hand-loomed products store we found for me a lovely sleeveless, woven tunic, soft, with a simple neckline and design, timeless, perfect for wearing over a pair of gray cotton 3/4 length pants.

It was just the color of pale tea.

Summer Meadow

I began dyeing canvas in East Cambridge. I was back from England, flat broke, and living with a friend from college in the only part of town near a subway we could afford. The small brick stores and narrow streets of East Cambridge reminded me of Europe. My neighborhood of Sixth Street had a fruit vendor, a couple of bakers, a fish store, liquor store, furniture store and printer. Of the bakers, the one we preferred was Santoro's, a small Italian bakery that made fresh hot rounds of bread that we could run out at 7 in the morning and buy for breakfast.

In the window the flour-dusted bins showed what was fresh that day, anise cookies, different shaped loaves of bread, long Italian loaves and small round ones, rolls and sometimes squares of pizza. The best part of that store beside the bread was that the family was so happy. They were often in there laughing at 7 a.m. in the morning and they all took vacations together in Florida, closing the bakery for a week and coming back tanned and smiling. The neighborhood had a discount fabric store called Sew-Low that offered tables of remnants of printed cotton by the door, all folded neatly, as well as bolts of heavier, more expensive cloth, behind. I bought fabric there as inspiration for the patterns in my paintings or just spent time looking at all the colors, weaves and shapes. I bought canvas for my paintings from the art supplier's downtown. I accumulated so many scraps of canvas from the varied size paintings I was making, I decided to experiment and make something new. I made a series of soft stuffed blocks, which I sold as children's toys and used as still life material. I designed the perfect canvas bag to hold all my teaching supplies as I substitute taught, and I made squares of different shapes which I incorporated into my increasingly abstract paintings. All of these I dyed in pots on the top of my kitchen stove, the kitchen that was also dining room and living room, discovering that several packages of dye brought out a much deeper color, the color I was often looking for. The blocks I dyed yellow, purple, orange-red, blue. The bag I dyed a bright forest green, leaving it in longer and longer until the thick canvas began to take on the color of rich, healthy grass. I draped it under the faucet of the bathtub and rinsed it over and over again until the last green washed out, then hung it on the back porch railing to dry in the sun. The individual squares I dyed red, and yellow, and blue. They found their place in aerial view still life paintings with checked tablecloths, aiding the play of space and color, the figure/ground of my paintings.

When I moved to Ithaca to spend the summer with a friend I adapted the dyeing to the weather and the style of living in that college country town. In the summer we would buy white cotton thin-strapped undershirts and let them soak in the bright pink and orange and blue-green boxed dyes, poking them with chopsticks and wooden spoons until they were pure, unadulterated intense statements. Then when they were rinsed and dried we would wear them while walking down the street in the bright intense heat of the summer, our small young bodies thrilling at the skimpiness of what we were wearing. We wore blue jeans and long cotton skirts and on our feet the black cotton and cardboard cheap old ladies sandals from Woolworth's: that was our look. We were artists. In the winter it was leotards and jeans and old fur coats or overcoats, and as much color as we could load onto our paintbrushes, palettes and canvases. We were proto-Matisse only we were in Upstate New York, not the Cote d'Azur.

When I was a student there, the last year when I lived off campus, my apartment was on one of the steep streets which tip downhill from college town to the flats, just down the way from the Summer Meadow Herb Shop. The shop was tiny and housed on the corner of Eddy and Buffalo Streets. Probably not more than a basement room, which some landlord decided to cash in on, it opened right up from the sidewalk and was lined with simple wooden shelves, and had a table on one side. Therese ran it, and I got to know her as I stopped by often on my way back from the art building on campus to try out the fresh granola, hand-packaged, and the dark aromatic lapsang souchong tea which I would steep in my kitchen and linger over in the mornings. Lining the walls were large jar after jar of tousled herbs that looked as if they'd just been gathered from a summer meadow. I would take them down one by one, and the teas, to smell and sometimes buy an ounce or two to sample. Fresh dried clover tops, mate, cinnamon sticks, red and dusty, curled like old manuscripts. Fresh black pepper, dark green catnip, spearmint, peppermint. Saffron in small plastic vials, though I didn't know what it was used for yet, dark yellow turmeric and mustard. The small darkish room was resonant with smells as I opened the door, heard the small bell jangle and began to wander around. Later Summer Meadow moved up and across the street and its early, guttural, earthy persona as the corner store was gone.

As I was leaving Ithaca for the last time, and it was a hard place to leave, I went thrift shopping to my favorite places downtown, bought bright flowered shirts and long solid color skirts and other shirts which I dyed—fuchsia, dark blue, peacock blue, as if stocking up on good color karma for whatever came next in my life.

Red Wine

Living in a Muslim country, now, we can only get wine from the U.S. Commissary: bottles of Australian red, some good Chianti, port. We don't drink much. I buy a set of very small glass tumblers at Newmarket and we take turns setting the table with them and pouring a small tumbler full of red wine at dinner. The glass just fits in your hand and I have noticed both of us cupping it thus, and enjoying taking a sip as we look out the window or work out a thought in our minds, to tell each other. Although I am not supposed to drink much alcohol, there is the opposing pull of how good it is for the heart, and the social aspect, too, and appreciation of the unique taste, especially when eating pasta.

When living in Somerville I used to buy a particular bottle of red wine at the small, gritty liquor store down the street, a Portuguese red wine that was only $2.80 a bottle. One bottle lasted a long time as it was only there for the sipping. When I came back from doing The Route, my artist's night job once a week delivering multiple real estate listing books, for which I got $75 a week, a fortune in those days, it would be 2:30-3 in the morning. I was both tired and wound up from running around since 6 pm in my overalls delivering telephone book sized packages. I would come into the brick former newspaper building, which had been turned into artist spaces on Walnut Street and settle onto the old couch in the kitchen and pour myself a glass of the Portuguese wine. I would sip the half glass of wine in the dark quiet of that grimy neighborhood, before the trucks, newspaper boys and buses started up in the early morning. That was before anyone was writing about red wine's medicinal properties. It was just something people knew.

I would also use the wine to tide me over on particularly hard days when my body ached and my emotions were driving me mad. One month I yelled to Jon who was sharing the studio—please—will you buy me a bottle of that Portuguese wine, I am beside myself. He ran down the street and was back with it, pouring me a glass so my clenched insides could begin to relax.

When I was still painting the red paintings (tablecloths covered with colorful odd objects and utensils, squares of color and sometimes pictures or cards)

I would also make red meals for my friends: lasagna and a huge salad with tomatoes and garlic bread and of course red wine.

Back on Sixth Street in Cambridge, just returned from England, I dreamt of myself in a red dress with a silver necklace and so one day I went out to Sew Low, bought two yards of unbleached muslin, and dyed the cloth a bright, orange-red. Then I sewed the dress by hand, using the scooped neck shirt of a friend as a pattern to cut by, only making it longer. I wore a silver necklace with interlocking circles that I had found at a thrift store. It was the first time I had worn red, having always been a light blue and purple sort of person. The red seemed to coincide with a time in my life when I was becoming far more expressive, more assertive; also more sensual. After awhile I didn't need to paint red, or necessarily wear red or even cook red. I just was that way.

Now at *Aranya* I watch the red madder being dissolved into a vat of boiling water. Madder is a root, found here. The only other way that I know red madder is through having been a painter all these years. One of the most beautiful tubes of paint, which is very expensive, is a deep, brilliant red color called rose madder. I use this color sparingly but it has the character of a brilliant, even dazzling sweetness. I am thinking of those roses in June in my childhood that seemed impossibly bright and lush, leaning heavily off the lattice where they had been growing and growing for years. Alizarin crimson is the generic color that has replaced madder. Its strength is as mixing color or a staining color. It mixes well with viridian green or ultramarine blue, to form an expressive dark. Much cheaper than madder, it is one of the first reds you encounter as a beginning painter.

Recently, I bought myself a red *shalwar kameez* in a new craft market not far from where we live. It has taken many months for me to figure out the kinds of fabrics I like here: the thick woven cloth of the Hill Tribes versus the light, filmy cloth of the cheaper clothes that sell in stores near D.I.T.2. The hand printed cottons vs. the repetitive machine made designs. In the beginning I visited a small cloth store, which had a very narrow selection, pointed at the two fabrics I would like, a grey and blue small pattern, interesting also in reverse, and a brilliant red with gold edging, bold and dramatic. The red yardage was short; I had to pick another kind, a gauzy blue and lavender. A tailor came to my house to take my measurements, and several days later delivered the garments to me: the dress-like *kameez,* with slits running high on the leg, and the very baggy pants, called *shalwar,* that reminded me of bloomers. The *orna,* the long scarf draped over the shoulders and bust, with long ends hanging down the back. For months I wore these two sets of clothing, until the colors in the market and the dress stores began to separate and resolve, and I began to know what I liked. When I saw the red *shalwar kameez,* just my size, on the rack of the new store, I was instantly taken with its simplicity and strength. A brilliant red, with small dots of gold sewn amidst a light pattern; red pants; and a long full scarf edged in gold color, forming a simple bright line of light when I threw it over my shoulders.

Now red is a color I choose with care unless I am being carried away by desire. Perhaps that's it: it's the color of desire. I remember a particular pair of red clogs I bought in London when I was a student, which showed up in a still life painting, juxtaposed with a postcard of a street in Amsterdam. Those were independence shoes, liberation shoes, rewriting the childhood fairy tale of the red shoes, which, once desired, can never be taken off. The girl in the tale, prone to dancing and other forms of high living, cannot stop dancing in the red shoes: finally she is forced to have her feet cut off, or die. In London on my own I discovered the color red, and still cherish its brilliant, vibrant warmth, healing the heart and infusing life with passion.

Indigo

I go over to *Aranya* to buy a vest and look around, wearing a printed *shalwar kameez,* made right here. Ruby is in her office. I sit down opposite her and talk about the business and whether I can get some more dye from her to use in my drawings. She is eager to have artists try the different colors, maybe use them permanently, and she is also talking with other people who may create factories and manufacture the dyes in larger quantities. I find her lack of competitiveness with others in this field refreshing. She is so clearly identified with the work and wanting it to take off in Bangladesh, not with her own personal success, although she confides in me that she thinks her store and its designs will always be unique. I am inclined to agree. There is artistry in everything that is done here. She wants to know how I have got on with my drawings and I tell her it took awhile to find the right paper, but now I am happily experimenting, only I have run out of dye. The indigo grew mold after awhile but I used the iron and molasses until it was done and I'm interested in trying the warmer colors. We talk about how these dyes would store over time in jars and how permanent the dyes might be. Then I get up to go as she is busy and we agree I will call soon to see if there is any dye paste left over, and come with my small plastic containers and get some.

As I sweep past the racks of multicolor clothing my eye picks out the blues of indigo and I have to stop again to study the designs and patterns on one particular rack. I know I will come back soon and get one of these blue patterned garments.

According to a flier promoting new indigo production, the indigo industry was begun in the 19th century in Bengal with British colonials and Bengalis working together. Bengal indigo became the finest and highest priced indigo on the Western market. But landowners began to force local peasants to cultivate indigo, to increase their profit. The bloody "Indigo Revolt" resulted from this coercion in 1859-60. [1]

There were actually three mutinies, but the second one in 1859 involved the most people and had the greatest effect. Beginning in 1800 there were scattered revolts and burning of *nilkuthis*, or blue houses. But in 1859 six million peasants got involved, burning, looting, fighting and boycotting. On the plantations, peasants burned indigo crops, killed planters and closed down the *nilkuthis*, using few guns but many locally made weapons. Women also participated in the fighting. The revolt spread throughout Bengal. The third revolt began in 1889, in the northern *Jessore* region of Bangladesh. At the end of the revolts peasants won their independence from the planters, choosing what would be cultivated, and growing it themselves. By 1897 much cheaper, synthetic indigo was developed, which dried up both the trade and production of natural indigo. [2]

Whenever I hear about indigo it seems shadowed, and I see now this has to do with the colonial oppression associated with its cultivation and sale. But in America it has another connotation—as I learned when staying on one of the sea islands in Georgia many years ago. I became enchanted with the landscape and the culture of the rural south, and also delved into its history. Along with cotton, indigo was cultivated, using the labor of African-American slaves before the Civil War. The very fortune of indigo was tied up with forced, indigent labor. [3]

I stayed on Ossabaw Island, which at the time was owned by one woman, Eleanor West, in whose family the island had remained for years. She had created two retreat centers for artists, writers and naturalists on the island. Old slave cabins and causeways built by forced labor and overgrown fields dotted the landscape. The roads were covered with crushed white oyster shell and the cluster of cabins also impressed with the broken white bits of marine life. Different areas of the island seemed to exude particular qualities. The road past the abandoned cabins that led to a causeway and eventually to an abandoned boat had a sleepy, silent character. The buildings themselves seemed dark and inviolate. The island was as mysterious as it had once been well known—with plantations and a southern pink stucco mansion and the work force, which transformed the land from its once wild state to cultivated fields. Now it was wild again, and largely unused. Except for a network of dirt and shell roads which webbed the island and the two artist retreats, one centered around the old mansion, one in the clearing of an old lumber mill, the miles and miles of land and tidal river and bay and beach were completely uninhabited by people. Wild animals and snakes created a natural Eden. We discovered each of the paths and roads had their own character too. Some were light and as it were blessed by the bowing cabbage palms growing every which way in the semi-tropical jungle, others strangely twisted feeling and troubling, where you sensed that a tragic act had occurred, a hanging or a murder. One road seemed to carry a chill.

It seemed stopped in time, with the same wild pigs running away when you approached and a twisted rope like a noose hanging high in a tree.

I came to the island first in an interlude between jobs, from a northern February to a southern beginning spring. I fell in love with the multi-layered island with its dark-and-light history, and huge live oaks casting shade. I knew I'd be back. I returned two years later with Michael, whom I would later marry and have a child with. We stayed at the more alternative retreat this time, called, appropriately, Genesis, and took long exploratory hikes down all the roads, and trips down many of the rivers, crossing the island, camping on the beaches and canoeing out to other small islands at high tide. It was during this time that I heard more of the indigo trade and sensed the ruins of a former life and society on the island.

Indigo blue had a magic ring to it, like gold spun out of air or the mysterious cure to a deadly disease given by a mythical figure. Perhaps this was because no one really used the word anymore. There was the British architect Inigo Jones, whose work and name cropped up in architectural history, and the unusual couple who thought of giving their child an exotic name, Indigo. But the blues which came to replace indigo had none of its troubled history: ultramarine, cobalt, cerulean, phthalo, cyan, sky. Indigo was linked with the history of slaves and, like other elements of its own checkered history, America would rather not touch it. The story of indigo was written in the records of the rural south in heat and hand-picking and misery; the value of it was so great at one time it was nicknamed "King Indigo."[4]

Indigo began to be cultivated in South Carolina in the 1740's, and in Georgia in the 1750's. African slaves may have brought it, ironically, as Africa itself had indigo fields and cultivation. Indigo dyes cotton and wool, and is made into a paste which is treated with an alkaline reducing agent. A chemical reaction turns the paste yellow and makes it water-soluble.[5] Britain supported indigo production in the colonies—they needed a source of blue for various cloth productions and paid well for the dye, although the American revolution ended these payments in the 1770's. "Indigo" was the Spanish word for the English word India, suggesting India's early development of the plant, where it grows most profusely. A colonial planter named Elizabeth Pinckney was said to have been an important figure in the development of indigo as an export in the New World. She took over the management of her father's plantation, and must have realized the demand for indigo was great in England and decided to try growing the plant. She was born in Antigua, then a British colony in the West Indies, and moved near Charleston, SC when her father inherited three plantations there. The whole industry came to a close after the Civil War in 1865, with the end of slave labor in the South. [6]

One of the drawings I do with indigo is intense and dark. Another artist who I met at *Aranya* says it looks like a monsoon. I am reminded of the change that took place in my work on Ossabaw—how I went to the island a still life painter and came home painting landscape. Stunned by the moving energies of the island, I started venturing out into its terrain, eventually rigging up a plastic milk crate full of watercolor supplies which I would balance on the seat of an old bicycle, pushing it along by hand. I would peer into the jungle that lined the road and stop when I saw something I wanted to paint. Sitting on the milk crate now turned over and made into a seat, I began a small, wet watercolor. In this way I studied palm bark and leaf growth, tidal rivers, trees stemming from pools left by the rain, and shells. To paint with the rhythms of the island I abandoned my cautious approach to painting, broadly flooding the paper with water, working into the paper wet, letting the images remain partially abstract. It was a leap in my painting that pushed me in the direction of Nature abstraction, and started a long series of abstract paintings continuing to the present day. In the indigo drawing I see the same powerful energies of Ossabaw, only now I am in South Asia, experiencing the land and weather of a tropical delta.

RECIPE FOR INDIGO DYE

Garbage can, 10-gallon—for indigo vat
Tall jug or jam jar—for stock solution
3 cups salt
1 ½ cups caustic soda dissolved in water
1 ½ cups sodium hydrosulphite
2 cups indigo grains
2 pints water (take out some to dissolve caustic soda)

The dye is made in two parts, the stock liquid and the vat. Each should be made in a tall straight-sided vessel so that as little liquid as possible is exposed to the air when oxidation takes place. When the dye is made one should try to keep the vat warm and at an even temperature between 60F (16 C) and 100F (38C). An electric element can be used or the vat can be placed next to a radiator enveloped in cloth or blankets. The vat will not work if it gets too cold or too hot.

VAT

Fill the garbage can with warm water to about 4 inches below the rim. Sprinkle 2 teaspoonfuls of sodium hydrosulphite over the surface while gently stirring, before mixing into the stock solution.

STOCK

Add the ingredients in the following order, stirring while adding each:

2 cups salt
1 pint water
1 cup caustic soda, dissolved in water
2 cups indigo grains
1 cup salt
1 ½ cups sodium hydrosulphite
Remaining water

Leave this mixture standing for 5 to 10 minutes, after which there should be a dark iridescent scum on top with a clear yellow liquid beneath. Lower the jug of stock solution into the vat so that it can be poured without splashing, to avoid taking air into the vat. Stir gently with a long stick. A blue scum will quickly form on top of the water but the liquid below should be clear yellow.

USING AN INDIGO VAT

Wet the material to be dyed in detergent (2 Tablespoonfuls of detergent to one gallon water) and squeeze out surplus liquid. Carefully lower the cotton into the vat without splashing. It should be at least 2 inches below the scum, so suspend by string, fastened with safety pins. If necessary the ends can be weighted down with bags of stones or marbles so that the cloth does not float. Leave in the vat for 1 ½ minutes, then gently pull out and allow to drip over a bucket, sink, bath or newspapers, but not over the vat. It will come out yellow and gradually turn from green to blue. Allow it to air for three minutes or so until it is quite blue, then dip it a second time for one minute. It will emerge again as yellow and turn blue in the air. Allow it to dry, and wash thoroughly. The two dips will produce a deep rich blue color but if a darker shade is desired, dip several more times. A considerable amount of surplus dye will wash out. The cloth should be rinsed many times until the water is clear. A piece of cloth dyed in indigo can be safely washed and even boiled with white cloth without affecting it. [7]

Eucalyptus

We were trying to get to the Anza Borrego desert, leaving New England during a particularly cold and snowy winter, only to find that San Diego was being hit by storms and rain. Heading for the desert, we left the city, heading high and east, a huge highway exposed. The rain turned to snow and the road was covered with slush. We turned north anyway, and made our way towards the western entrance of the desert country. But when we reached a state park, before the road became truly mountainous and would then begin descending through circuitous routes to the desert floor, we were driving through three inches of snow and thought we'd better stop and ask what the "road closed—go no further" signs were about. I managed to find a park ranger in the silent and empty place, who advised me, indeed, that we should turn around and not try to go any further. So dejectedly we turned around and made our way back to Pacific Beach, where we found our old motel, this time renting a room that had a glimpse of the sea, and two tall palm trees. We watched the Olympics on TV, my son tracing highways with his trucks on the bedspread; read novels, went out for walks by the ocean, ate take-out Japanese food and sat in the late afternoon at outdoor cafes that looked out to the gray sand and water. I collected many big, conical shells with glassy, silver interiors that had been deposited as a group at low tide outside our motel, to use as inspiration for my paintings.

And then we decided to try again. This time we took the northern route which had many fewer exposed roads, and not understanding distances very well on the California map, made reservations in Escondido, which looked, abstractly, like the last stop before you turned right, east, and started heading for the desert. The one thing I knew we had to have in our hotel room was a fireplace. Although we were in Southern California, it was damp and cool so far. I pictured us holed up in a lovely cabin with a cheerful fireplace as we read and talked through the evening, before we headed out to the desert.

We were on a huge double lane highway when I realized we were near our destination. To my shock the motel I had so carefully chosen was right on the main road. We turned off, and then made another right into a kind of parking lot compound that included a Wagon Wheel restaurant, a long, low-slung unexceptional-looking motel, a pool surrounded by wire mesh fencing, and several trees. The Hispanic proprietress showed us the end unit I had reserved on the phone, and I couldn't help being disappointed by the old and somewhat cheap fittings, the spaces that had a lingering tobacco smell, the thin towels. The pool was unheated and the weather still quite cold, so we didn't swim, but certain qualities made it a resonant and an almost healing stay. The room did indeed have a fireplace—it was huge and wall-sized, lined in brick, blackened with use, just next to the large double bed. There was plenty of firewood. The proprietress brought us a pile and we helped ourselves to the rest, just outside the door. We arrived in late afternoon, and by early evening, when it was dusk, we started a fire. We'd brought a picnic supper, which we ate on the bed—hard boiled eggs, pickles, soft cheese, crackers, olives and some wine, and as the evening grew dark the three of us, Michael, my son and I settled in with books and drawing paper, and the fire. When it was time for bed we stoked the fire and put the screen around it. We tucked my son into the foldout bed. The air was filled with a smoky warmth that tingled and seemed to dry and cleanse the inside of one's head. The flames from the fire made moving bright shadows across the walls. I don't remember sleeping deeply. The night was laced with the energy, roaring , brightness, and woody fragrance of the fire. It was like a purging of past annoyances, a cleansing of the mind, an unexpected, roaringly funny joke. We woke, delighted and rested. As well as the fire, the row of trees which grew between us and the highway had a special effect. They were eucalyptus, and they moved and scratched against the windows all night in the most keenly alive way. Opening the door the smell was dry, tangy, timeless and dusty, making your head spin. The traffic was forgotten, the dinginess of the surroundings gone.

In the morning we rose and walked across the blacktop to the Wagon Wheel restaurant, which turned out to be a big greasy spoon with huge breakfasts of eggs and home fries, toast and coffee. Despite the roughness and tackiness of the general surroundings and the disappointing first view, we had found our own small neighborhood for a night that gave back to us unexpected pleasure. What appeared to be a dump was full of potential meaning. The eucalyptus and the fire, the proximity of a simple breakfast and the direct, unaffected sense of place became a healing moment in time.

And then we hopped in the car and headed for the desert, sunny day at last, passing beautiful, crisp pines and small towns, some still with a dusting of mountain snow. We descended to the desert floor, to spend the next few days crushing sage between our fingers, going for walks and studying the pebbly gray hills that rose straight up.

What I didn't know at the time was that in two months I would be struggling with cancer. In retrospect, the fatigue I was feeling probably had a cause. We were traveling on the edge of a ravine and didn't know it. Like the weather that was so uncooperative, that kept us from getting to our goal of the desert, there was a shadowy force in wait, something we couldn't control.

Finding enlightenment in the ordinary, we were turning straw into gold. In those curious off the beaten track experiences, we were grabbing the present, like the monk in the tale on the edge of a cliff who is chased by a tiger, but reaches to enjoy the strawberry growing off the side of the cliff. We were studying the hills, and the desert, and smelling eucalyptus.

Frangipani

On one end of the island of Bermuda is an old perfume factory. It sits on a piece of land studded with palm trees and tropical growth, orchids among them, wound with paths that lead next door to a plant nursery. In the small white buildings, which are a kind of museum as well as a shop, the visitor can see how perfumes are extracted from the various flowers of the island. In the last room one can buy diminutive vials of these extracts, which are then packaged in tiny white rectangular prism boxes. The place feels eccentric, old and handmade. There are benches outside under arbors of yellow orchids and small stone patios to linger on and view the flowers. Not being a perfume person myself, I went with my friend Ada who loves jewelry, perfume and flowers. This was years before Bangladesh; the rest of my family was occupied. We were running away from winter for a four-day excursion. Bermuda was the closest place to go that had warmth and foliage—it was only two hours away. I would get ideas for paintings. We met at 7 in the morning underground at Harvard Square Station—in the damp and greasy late winter morning, feet cold and sliding through snow debris on the platform brought by travelers. Want to go to Bermuda? I asked Ada and we both laughed. It was so strange to be meeting here. Michael was on one of his trips to Asia and my son was with my parents for a few days. We had decided we deserved this trip—even needed it.

But Bermuda was more cultivated than we thought, and we found ourselves seeking out off-the-beaten-track places if we could find them, lingering over a yellow bird seen outside the window in the early morning, an empty beach outside a nearby deserted hotel (it was off-season), a small shop that sold English china I had seen years ago in London. We rode the ferry, we walked the old railroad line and picked up found objects. But the best place we found—maybe because of the land it was on, still somewhat wild-seeming and offering a path, which is always a good sign, was the perfume factory.

Ada is my older friend who works at the one of the museums at Harvard. She had lured me to Bermuda by her talk of pink beaches and turquoise sea. She was a sensual, visual, visceral lover of life who went into rhapsodies describing minute details of her time spent working in Iran, years ago. The blue glass vial that a woman kept to collect her tears while her husband was away. A certain tree in Cambridge that was like the canopied trees in Iran. Her garden lilacs in the spring. Fountains. A rare painting. Once we met at the Christian Science building in Boston just to walk around the reflecting pool, then laughed hysterically all the way home on the bus. Another time we met at her apartment in Harvard Square on the Fourth of July and walked the length of Mass. Avenue from the Square to M.I.T., the street closed for the evening, feeling we were in another culture, with people hanging out at cafes and strolling leisurely down the middle of the street. We stopped at a small Indian restaurant for dinner and lost track of time and stumbled out into the darkened night, walking south to watch the fire works as the crowd got denser and denser.

She was my muse at times, showing up on a Sunday morning in her red floppy hat to sit on the back deck and drink coffee with me. I could always write or paint better once she'd been there. I visited her in her small apartment above Harvard Square and we went through a phase of drinking Linden Tea and eating madeleines, which she had baked, in the mode of Proust, small shell-fluted cakes which we dipped into the pale yellow tea. She was always there in her office when I visited Harvard, and had a hand picked choice of painting for me to see. She knew the specials at the museum gift shop. She took me out to lunch. Now in Bermuda we explored the off-sites, and she bought frangipani perfume, which she brought back in small, white rectangular boxes.

Bay

The house we have moved to in Bangladesh has a huge old bay tree in its yard—as well as palms and palmettos, banana trees and huge hibiscus. I was drawn to the property by the old elaborate scrolled white gate, and the sense of trees guarding the house—and of course the small hand-lettered "to Let" sign hanging on the gate. I stopped the car and knocked on the gate, and eventually a guard with a Muslim cap and *lungi* came to open it, and I walked inside. The big bay tree was at the other side of the property, hanging deeply over the yard and making a dark pool of shade of the corner. I walked over to it, and have many times since, to pull off a leaf and smell. Yes—the very kind used in cooking. Peni says you can make a kind of tea with the leaves that is very good for you, and I have it in mind to do this. Razia says you can see the bay branches drying on the side of the road in northern Bangladesh. I wandered slowly around the yard—down a small path that led to the water faucet, worn orange through the grass, and around the outer perimeter of the wall, where big banana leaves large as umbrellas stretched every which way like mad sun dials. There was a haze of light pink where wild rose bushes spread, and points of brilliant red where the hibiscus bloomed.

From inside, the big picture windows of the living room looked out to this scene—a wild collage of green and yellow and red with soft palm parallel strokes falling down from the sky. It was totally appealing, totally alive and inspiring. I resolved to bring Michael back to see it. In time we rented it, leaving our dark, narrow house with no yard for the sunny openness of this place. All eyes, all windows.

Ada had been the first to introduce me to bay. She brought a Christmas gift to me at my apartment in East Cambridge—a small orange clay pot with mixed herbs, 'pot au feu' snuggly filling the inside. I knew so little about herbs, then, and was not much of a cook. I knew teas, and had recently been initiated into knowledge of grains and rices, but knew very little about growing things or the finer distinctions between herbs. She gave me, to an observer's eye, a pot of green flakes, with many unusual smells.

On my fire escape, just outside my studio bedroom I had planted tomatoes in found metal buckets. A fig tree grew below our porch, in the Italian landlady's yard. One yard away a healthy apple tree exploded in red fruit. My apartment mate grew cannabis behind her bedroom door with a grow light and fried it one leaf at a time in an iron skillet on the stove.

This clay pot of herbs stayed for a long time on the shelf above the stove. I may have tried using the herbs once, but mostly they represented exoticism, the unknown, and nature used and understood, if so much could be contained in a small pot.

Ada also lent me her unusual choice of reading at the time, *Witch Doctor's Apprentice,* a terrible title but a fascinating look at the use—and deliberate loss—of samples of medicinal herbs in the laboratories of a US drug company. She and I seemed to share a sympathetic strand of interest on the use of plants, the discovery of healing methods outside the realm of traditional medicine, and the value of the journey and adventure, in life, as learning tool, a point of inspiration and source of knowledge.

When we took the #69 bus to Sixth Street, we hopped off in the chill of winter slush and crossed the narrow streets to dash down the alleyway, roof dripping on our heads, and climb the back stairs and open the kitchen door. When we met among the venerable oaks of Harvard Square and went off to find some good coffee, we discussed this book, a real sleeper I have never heard anyone else mention, and what it might mean—for our lives, for the world. Might there be a language of plants that none of us in the West knows about? Might we trust our intuition about this?

Now I go out in the yard and make contact with the bay, pulling a branch down to get a leaf and dry some to use as herbs. In my cooking I have graduated to a wide range of herbs through sampling and through living with Michael, a terrific Asian cook. What seemed mysterious to me now feels familiar, although I am always learning about new herbs, and new uses for old ones. Recently a Pakistani friend gave me tea with cardamom, and in our own house when Paul makes tea he often adds cinnamon or clove, so that it takes on the quality of that high-priced *chai* we buy in the States, only here it is quite ordinary tea. On the corners of the two streets where we live are tea stalls, small impromptu shacks, *cha dokar,* with a single wood counter hosting a jar of cookies and *samosas,* and cups of tea. The Bangladeshis who do work for us buy their tea there, several cups a day, and I have also noticed it is a small gathering place for men and a watching place as the theatre of the street, which includes our own comings and goings, unfolds.

RECIPE FOR SOUTH ASIAN TEA
Tea is made by boiling water in a pan on the stove, then throwing a tablespoon of rough cut tea into the boiling water. After about a minute, pour in milk until the liquid is the color of coffee, and let it boil a bit longer. Meanwhile put a spoonful of sugar into the waiting teacups. The milky tea is poured through a tea strainer or sieve into the waiting cups.

BAY TEA
1) Add several bay leaves, dried, to the boiling tea before adding milk. Strain leaves and tea together.

Or,

2) Fill tea ball or strainer with dried bay leaves. Pour boiling water over leaves. Steep for 5 minutes. A delicate alternative to well known packaged herbal teas.

Durian and Jackfruit

On the coast outside of Bangkok the streets are full with people selling Durian, a crusty fruit I have been warned about and which is much maligned by some. On the market tables the fruit lies opened yet still in its shell, pale beige fleshy-looking compartments soft to the touch. Gingerly, I try a piece and find it similar to sweet potato—not offensive, but not a favorite either. In the hotel our friends keep their durien out on the balcony where it matures— Michael swears he can smell it even in our room. Jackfruit is the other common fruit of South Asia. I see it growing everywhere in Dhaka now that it is the rainy, hot season. It hangs off trees, huge, green and pregnant, waiting to be picked. The jackfruit leaves are large and glossy. A friend says she has tried jackfruit curry, with the seed left in. I wonder, driving by the sea of jackfruit trees, who eats all this fruit and what will become of it if it is not picked.

On the street near the American Club is the fruit vendor who sells both local and imported fruit. He raises his prices for Americans. Sometimes I give in to the familiar look and smell of grapes and peaches and avocados and buy some despite their price. Avocados could grow here—Bangladesh has the perfect climate, but no one has tried it yet in bulk. In a country such as this one eats by the season: in summer the vegetable selection is very narrow: carrots, cucumbers, tomatoes, spinach but in the winter, zucchini, broccoli, red spinach, okra. How different it is not to get every vegetable one desires or craves, but to live by the way the land produces.

Next Fall, likely our last in Dhaka, I will plant herbs and greens, in order to live by the garden.

Painting With Natural Dyes

We had been asked to bring several yards of thin smooth white silk, washed, boiled on the stove and ironed to *Aranya*. To buy silk for the painting project I go to *Rajashahi* Silk, on *Kemal Attaturk* Avenue, not far from where we live. Bolts and bolts of stingingly bright silks fill the glass cases behind the counters; folded lengths of raw silk lie underneath the counter itself, which one has to stoop to see. I realize the second time I come to the shop, the selection is always changing. One day the colors beneath the counter are all natural beiges and yellows and dull greens. Another time they are brilliant blues and yellows. I ask for a smooth, thin silk, and am shown a delicate white bolt selling for 150 *Taka* a yard, about three dollars. I buy four yards of it, and take it back to prepare for dyeing. It needs to be boiled, with detergent, for two hours, then rinsed, hung up to dry, and ironed. This removes any starch or sizing from the fabric and allows the color to take.

Four artists, Alka, Arham, Rokeya and myself converge on *Aranya* to try painting with dyes. Alka, who is Indian, and Rokeya and Arham, who are Bangladeshi, like myself have established careers as artists. I am the first to arrive, and have a bemused half hour sitting in the office while people come and go; I am served rich dark tea with milk and sugar off a small tray. I notice a locked bookcase with volumes pertaining to dyes, wild flowers, native plants, printing, and many small plastic bags filled with samples of colored dyes on fabric. I tend to forget what Asian time is. With my western sensibilities, I arrive punctually, only to find that my Asian counterparts arrive gradually.

When the other artists come, we all move into the workroom with long tables and overhead fans . Ruby changes the cloths on the padded surfaces, to make sure we don't pick up patterns and colors from previous dyeings, and we each spread out various squares of silk, pinning the corners to the table cloths with straight pins, to avoid wrinkling and movement which might be caused by the overhead fans. Ruby has specially prepared some silk so we can paint with red. It requires a particular mordant added to the fabric, which I will learn about eventually, so when it is steamed and sunned, the color remains true. She sets out five bowls of color: red, black, orange, indigo and yellow. Later I will learn the *Bangla* names of these plant dyes. Alka loans me brushes, which I forgot to bring, and we begin to paint.

The bowls of thick color resemble bowls of curry. They are slightly syrupy, thick, but can be diluted with water. As I begin to paint I am reminded of the process of my ink drawings, where I dip my pen into small dishes of ink again and again. Also watercolors, where I am constantly adding water to achieve soft washes and delicate color nuances. I dip the brush in the dish of dye, sensing a slight difference as it touches silk, which is extremely porous, compared to paper. Depending on the amount of water on the brush, or on the silk, a distinct, dark mark appears, or quickly softens into a larger haze with a darker streak at its middle. I use a lot of indigo, enjoying the slightly sharp aroma it carries, and black. But I also dip frequently in the yellow, and mix a kind of brown from orange and blue, which I twirl on the paper towel next to my painting, and then added as deepening accent.

I do four smallish paintings on silk, standing to gaze long at the marks and pattern as I am nearly done, walking back and forth, seeing if a spot calls to me or needs strengthening or enhancing in some way. I notice we all get quiet as we get involved in the painting, being lost in that particular kind of intense reverie that happens when one is deeply at work. After a while we all start looking at each other's work, giving feedback and saying what we see there. And then the paintings are finished, at least for now, and I decide to walk home. We don't live very far away—all I have to do is cross the bridge from *Banani* to *Gulshan,* and the very busy street, and walk west down the increasingly house-lined streets until I turn down the road which runs along the far corner of the property. I quietly walk the last few hundred yards, gazing at the palms and the people walking by—there are no cars on this small street. We all agreed to meet in about a week if there has been sun, required for fixing the color, to see the next stage of our dyed creations.

As it turns out I am the only one who calls to see if the fabric has hung in the sun long enough, and to come over to observe the process of steaming. At the end of the week my son and I are going home to America for a visit, and I really want to see the next step before I leave.

Ruby shows me how the silk pieces are put into plastic bags, tied once, enveloped again in another bag, and tied again securely. Then they are put in a huge drum of water boiling on a small gas burner. The drum is lined at the bottom and the top with burlap, to absorb the wet steam. It has two sections, something like a double boiler, the burlap and cloth in plastic bags in the upper portion, then covered with a lid. A large brick sits on top of the lid, to make sure the cover isn't pushed off by the steam. The fabric steams for two hours. I am shown how I can do this process at home, using a large stainless steel pot and ordinary kitchen plastic bags, string and cotton cloth.

The room the steaming takes place in is off the production room. Very small, with stone floors stained reddish orange from dyeing. The door opens to the small alleyway between buildings, where the rinsing, in low concrete troughs, takes place. All is much worn over time, worn implements, familiarity, vessels, a sense of use and practicality. Next to this is a very small room with a hotplate for making tea.

Ruby and I talk about the possibility of putting up a show of paintings using natural dye next October or November. Is that too soon, we wonder? Should it be combined with a show of indigo products? We agree it should, and we will leave it as October, but we can wait till November if necessary.

In getting involved in this process—the use of natural dyes and painting with them on cloth and paper—I feel as if I am participating in an interaction between nature and color. Nature grows the plants used for dyes, but through a process discovered and created by people, the essential elements of color, embedded in the plants' nature, become useful paints. By dipping in the paints, and making a stretched fabric, the signature of plants becomes known, at the same time the artist's does. A collaboration, of sorts.

In a folk tale about indigo which someone gave to me when I was on Ossabaw Island, in Georgia, the interaction and distance between human and natural worlds is so close, the heroine of the story is continually pulling off blue from the sky, in a magical way, to make into indigo paint. Through a sort of hubris-like experience (similar to the Greek play's hero or heroine who asserts himself too aggressively in relation to the Gods and is punished) the woman in the folktale, so involved in her dyeing, neglects a significant part of her life, her child. In the end this loss is a gain for all of humanity, a sacrifice of sorts. Thenceforth the indigo dyeing process becomes easily accessible and known, while the sky, the original source of indigo, retreats up to its current location. [1] As an artist one gives up the known, in order to continually enter into the unknown, to make new work, to create new forms. An awareness and wisdom come from working with new materials.

I remember traveling in Scotland for the first time and noticing that the colors of the heath—soft roses, greens, lavenders and yellows looked, altogether, like plaid. I learned later that heather creates a greeny yellow dye, as well as being a tea and a medicine.[2] And African *Kente* cloth also resembles the vivid colors of its vegetation. The Vegetable Dye Society color chart has also its parallel in the Bangladeshi countryside. Deep blues and greens of indigo, warm yellow radiance of grass and rice field, warm browns and oranges of tree trunks and all the in-between earth colors that make up this tropical delta environment. When I first had the chance to get out of Dhaka, and its urban sprawl, on a riverboat trip to *Kulna,* I was struck by the beauty of the low flat landscape with its intense green rice paddies contrasting with grey blue water and sky. We were painting in the colors of the land of Bangladesh that day on silk— like the other arm of a compass, its point firmly fixed in the landscape outside Dhaka, its other arm inscribing more than a circle.

I saw birds, flowers, energy patterns of nature in what we made. We were mirrors, reflecting back what came through our brushes. Perhaps we were meant to be painting with the products of the earth, like the first artists, cave painters and rock carvers, inscribers of eagle bones who used what was at hand to describe what they saw.

Basil

Basil has been with us since ancient times. It apparently came to the West via India and Southeast Asia, and Holy Basil, *tulsi,* is a sacred substance to the Hindus in Malaysia and India.[1] Holy basil is planted around temples to keep away mosquitoes.[2] It can be used as a dye, where the leaves are cut in summer or fall, simmered and strained. After long soaking, the dyed garment, a yellow-green, is dried in the sun.[3] In Italian cooking, it balances its brilliant green with the succulent sweet red of tomatoes.

For several seasons my partner and I had an "only basil" garden, but that is long after I discovered the uniqueness of pesto. I was living in Union Square, Somerville, in a studio building with four other artists. Jon and I shared a floor, which included living and work spaces. We were friends, passing each other in the shared kitchen/living area and doing each other favors as we could. I joined the Food Coop in Central Square, Cambridge, and made weekly runs there for good farm vegetables, fresh bread and dark, intense ginger muffins. The informality of the basement food store that was continually in the process of being re-stocked, and the unusual vegetarian and organic products you could buy there drew me despite the parking situation outside. I got out my Moosewood Cookbook and began making foods in a quantity and range of tastes that suited me. I had heard about pesto, but never tasted it. It was way before the vegetable cooking craze that made herbs and spices easily available. Despite being only a so-so cook, since the end of college I had been drawn to herbs, spices and vegetables as if to a grail. Some invisible thread, a sympathetic connection to nature, maybe, or a resonant strain of understanding as an artist of pure colors. The first book I bought which had to do with vegetarian appreciation and cooking was *Victory through Vegetables.* That and Helen and Scott Nearing's *Living the Good Life* were all that you could find in the bookstores on the East Coast in 1971. They were off the beaten track, but they drew me.

In my spare kitchen on Walnut Street, just off the studio, I decided, in one of the first meals I cooked for Michael and myself, to try pesto. I bought a fragrant armload of basil leaves and settled in to figure out what this food was. Not having a blender those pre-tech days, I chopped the basil by hand. It had a ragged, green slaw kind of texture, and I mixed it with the requisite olive oil, ground-by-hand walnuts, and grated Parmesan cheese. When Michael came over for dinner we sat in the former newspaper-office turned kitchen of the old studio building and ate green food: thin spaghetti mixed up with the beautiful green textured paste that is pesto. Later as I came to make it more often, and we had a blender to grind the ingredients, it turned into a rich mortar of grainy ingredients, nuts and basil and oil and cheese uniformly chopped. But that was long after we had started to grow basil ourselves. As we sat in the newspaper office-turned kitchen, Michael turned to the windows overlooking an overgrown yard, a tangle of trees and weeds towering over the short wire fence and imagined the *Barong,* mythical beast from the Balinese story of good and evil, whose mask he had hanging in his apartment, bursting out of the back lot, surprising us with his dancing gait and bulging eyes. There was always a shadow of Asia in our talks and in our imagination, as Michael had lived there and worked there for many years, and I felt drawn to its art and religion. The *Barong* fought the witch *Radha* in an endless contest between good and evil. The *Barong* represented the good forces in the universe. The Balinese continually acted out the overcoming of evil, in an effort to keep the balance of life intact.

When we lived together near Powderhouse Square in an upper apartment down the street from the park, one of Somerville's few, we had no yard ourselves to garden in. So our friends Annie and Mark up the street offered us a small plot of land on their property to cultivate as a garden. As we had become avid pesto eaters by this time, and had refined our methods so that we made lots of pesto each time and froze it in portion-sized containers, we decided to grow a "basil only" garden and go into production of pesto, no holds barred. We must have planted fifteen basil plants in our small plot, and came up the street often to water and weed them. The tops needed pinching as soon as the small white cones of flowers appeared and the plants needed full sun. By August we were making pesto. Our idea was to lay up enough to freeze so we could eat it all winter, and we did. This became a regular seasonal practice, and we seemed to take turns, first Michael, then I, making the huge batches and freezing fist-sized amounts of it to store. One summer we didn't have a garden, and Michael bought huge bouquets of it back from Faneuil Hall, what remained of Boston's huge outdoor market.

All winter long we ate pesto—a green alternative to looking at the snow and ice clouding the long New England winters. And waited for the first whiff of fresh basil, surely the sign that summer has really come, in June or July.

In Bangladesh we eat pesto but have no blender, so we return to the ragged slaw of basil cuttings mixed with walnuts and ground Parmesan cheese and olive oil. When we grow tired of the heaviness of Bangladeshi food we often revert to a big dish of pesto, with garlic bread and red wine and salad. Sitting in the darkening dining room with its heavy, long wood table and chairs, its hangings from Nepal and paintings from Bangladesh, photographs from Thailand, I am reminded of our beginnings, the small garden up the street in Somerville, the connections between time and places, and the ubiquitous herbs like basil.

Painting the Delta

I have gone over to *Aranya,* it's a Thursday afternoon when we're all supposed to be painting, but no one else is there of the group, just Moni and Ruby in the office. They're not even sure they have any paste. They're distracted by visa problems, and I'm distracted by terrorist bombings in the U.S.—but I ask if they have any old dyes I can take home. I've brought containers. I manage to come back with three—*catechu, indigo* and *shikorai.* Ruby says this will help them know how long the dyes will last, whether I can use them at the end of the week and still find the color good. I have come with a roll of paper and a fresh pad of paper, and so, since my driver just dropped me off, I wind my *orna* scarf around my shoulders, pack up my bag with dyes and head out to walk home. On the way I veer into the fabric room, bolts of silk and printed cotton and admire a new pattern, in soft, warm brown, a variant on fleur-de-lis. And I say hello to Mr. Hussain who is printing fabric on one of the long tables in the workroom. He goes back to work immediately, being rather silent in general.

Outside it is windy and the traffic is heavy. I am bundled up against the eyes of strangers and the calls from rickshaw drivers. I pass the lake and gratefully let my eyes drift out to the horizon of white buildings on the far shore. In the city to get a long view like this is like a balm to the eyes. It takes me nearly ten minutes to get across the road with the traffic, and then I turn in to road #50 which is muddy and filled with puddles, people and vehicles. I look down and the outer roll of my paper has been splattered by a car going by. In the mood I'm in I think—at least it's natural, and, not unlike the brown dye we've been using. I get such pleasure from simple walks here because I don't do them very often. To the club, to the Korean Restaurant, back from *Aranya.* I turn into my cross street and slow down—I had been marching fast down the street—and sashay down the small street with all the palms overhead. Green, moving, alive.

Later—we are waiting to find out if we'll be evacuated. The world situation is tense after the bombings in the U.S. Being in a Muslim country puts us at risk. There are many days when the embassy advises us not to go to downtown

Dhaka, and we can hear demonstrations from our house. We've turned Canadian for awhile, if anyone asks. The painting at *Aranya* has been suspended for now as Ruby and Moni and one of the other artists are all away. I have three containers of dyes in my studio, which they gave me last week before they left. I've used them once. As with much of my work, the best of it does not spring full-blown from first attempts. When I go back to look at what I did last Fall I really honestly find only one painting/drawing that I like and that I would want to show. It's the stormy set of marks in Indigo that looks like inner and outer weather. The paintings on silk I did at *Aranya* which I saw for the first time last week are disappointing—they seem to have bleached out and the subtle washes I thought I was getting when I added a lot of water to the dye are gone altogether. So I will have to work hard to have pieces to show in the Fall. Today I realize if I don't use the dye soon it will go bad—so I take out a stack of hand made banana and water hyacinth paper, cut each piece in half so I have a healthy amount, open the containers of dye, gather brushes, clear the table, bring out a container of water, and begin.

What with the heightened feelings of the last week, the new world situation, I find my emotions running strong, which is giving me an energy I haven't felt since before I was sick. I start dabbing and brushing and scraping and even scrubbing and before several hours are up—not even interrupted by lunch, which I have forgotten about—I have 16 drawings. I locate one of the three I did last week and add to it, using first *shikorai,* a light golden transparent color, *catechu,* a brownish creamy gold, then indigo to build up a patina. Doing 16 drawings of course does not in the end yield 16 good drawings. Even as I'm carried away by the thrill of being completely immersed in painting, I also know in some other part of myself that these may all be prototypes, and ideas—rehearsals, a form of doing research, for the work to come. If I am lucky, I will come out with maybe 2-3 or maybe only one resolved drawing after this afternoon. This kind of afternoon of plunging into the work and rapidly pulling drawings off like prints is so unusual: daily I plod through longer, more demanding paintings and intricate watercolors and far more concentrated ink drawings. But these today seem to be fueled by the situation the world is in, my sadness at the disaster in the U.S., my fears for all of us. The dyes were there, the paper in a stack, the brushes ready to use. Tonight or tomorrow I will come back to view what I have done, rip some drawings up and put up others to contemplate on my wall. I have a fear that if we are evacuated soon I won't get a chance to show this work in the Indigo show—and I want these marks to be included in *Aranya's* history and process.

There is strong anti-American sentiment since the terrorist attacks which may necessitate our leaving. I have a suitcase packed with essentials. Everything else will have to be left behind.

Looking over the dye drawings, I choose seven including the early dark blue *Delta*. I photograph these out under the bay tree, in case they get sold, and take them to *Aranya* to check with Ruby and Moni to see if they agree on my selection. There for the first time I get recipes for dye paste so I can make my own inks, and discuss with them the next steps in creating the show. Ruby likes the work, but says—what about these—these aren't indigo, pointing to four of the recent paintings. But I protest that indigo is there, the top layer, just affected by the warm light and dark beige underneath. The indigo turned a dark shadowy purple-grey on top of the other wet dyes. Two of the pieces are pure indigo—I had that left as I was furiously covering paper on that day I was so productive—they are huge dry brushstrokes with clumps of wet indigo ink at the base, they seem almost like Japanese ink drawings, which my work has already been compared to. Pure process, exciting in their ability to capture motion, a kind of natural weather or energy.

I find that this work on paper is carrying me back to the work I did before I was sick—it has a kind of energy I lost when I gave up traditional oil paints, and their dangerous solvents. Painting with oils was making me tired and weak, as my compromised immune system tried to cope with the effect of turpentine and paint thinner and linseed oil. But the thickness of the oils and their ability to stay wet, and be able to be scratched and textured, felt very much like the rough, textured surfaces of nature to me. Something about using these dyes made from nature, on paper made from nature, feels curiously right. I can't quite articulate it even in my mind right now, but the air and space coming through these drawings—not to mention the sense of hand made materials—feels authentic and real, similar to my invention years ago of a new way to draw, new tools to draw with. I dropped the conventional tools of drawing, that made me want to perfect whatever I did, and make it smooth, and chose to use sticks which I collected in the woods, and then shaped. I began by working outside, trying to capture on rough paper the marks and patterns of sunlight and shade that made up the tapestry before me.

Now I imagine using natural pigments and sticks to communicate, like ancient paintings on the sides of rock. Maybe we need these connections, in our overdeveloped world, between ancient artists who lived in Nature, and our own efforts. Once again I am humbled by my inability to necessarily grasp what I am doing until later, when, like a hologram, the meaning/image of my work will emerge out of thin air and the marks that were there all along will speak.

Morning Glory Blue

We always saved morning glory seeds from previous years and gardens. Despite their seeming annual nature, the flowers came back year after year to wind themselves around the tall posts of our second floor deck in the backyard. Mary, who lived downstairs, was the true gardener. She tended her small plot in the corner of the yard with loving care, producing a riot of color—and flower cuttings—year after year. Together, we planted many morning glory seeds around the posts, and watched as the bright green tendrils made their way up the slender wood steles. By midsummer they had undergone that burst of energy plants do, and multiplied in bright blue violet profusion in the mornings. Mary drew the vines together from posts about 6' apart, helping to create a halo of vine and flower about 5' off the ground, growing ever more abundant and thick as the summer days progressed.

By August there was a cloud of bright green leaf and small blue flower, unbelievably cheerful to confront when one wandered down with one's cup of coffee at 7:30 am, to survey the garden realm. Towards September it was cleanup and tend time. We pulled and raked and generally harnessed in all this wild green energy that had grown unchecked in the last 3 months. Someone—I forget who—pulled up the roots of the morning glory vines and cleared out the tangle underneath the deck. I was shocked—even a bit despairing, to see the green profusion go. Need one admit that summer was over, and the night air becoming cool? The tangled halo of flowers was still there, presumably to dry in a kind of arm and shoulders wreath.

But for several mornings as I made my way downstairs to the garden I encountered the uppermost parts of the morning glory vine still in bloom. Although there were no roots—no lower vine at all—and no connection to the ground, the morning glory flowers were blooming as if nothing had happened. It was as if the crown of blue and green had a life of its own—an inner life—that sustained it even as it was supposed to have died. None of us could explain it.

There was a paradoxical orange note, opposite to blue on the color wheel, in the garden, too. That same Fall we heard a commotion in the garden that met our garden corner to corner. The back two plots, directly behind the house and to the left (the one in question) were owned by an extended Italian family. Right behind us was an unbelievably well-organized and maintained vegetable and fruit garden, good enough to be in Italy itself (and Somerville was a little Italy). Periodically Tony, its owner, offered us dandelion greens and cucumbers from the garden. In the spring we watched him bring up the fig tree, buried horizontally in the earth. He also spent a considerable amount of time trying to persuade us to cut down our large old maple, in the far corner of our lot, just grazing his driveway and trying, as he said, to shade his garden. He had cut it back as far as he could in the part that overhung his property, so that it had a kind of triangular existence in our yard, its branches extending out in a right angle piece of space, over our land. He never did persuade us. We loved that tree.

Next door to him lived his brother Philip, older and stouter, who had the kind of yard you could rummage around in—sheds and shacks and a back vegetable patch not nearly as organized as his brother's. One Fall Saturday in question he and his grown up sons were harvesting squash and pumpkin. Pulling on one particular vine they discovered it had escaped the tangled darkness of the yard, and wound its way through the old tree-like lilacs in the corner of Mary's garden. Just below the level of tall flowers that filled the back of her crop, hung a perfect, medium-sized pumpkin. Airborne, it had matured in peace, in space, in a surrounding of flowers. Innocent, we had allowed it to grow in silence and anonymity. We thought we might perhaps have some air rights to this pumpkin, but Philip's sons stormed over, crossed our yard and, friendly but determined, broke the pumpkin from its stem. But not before Mary had taken a picture of it—which exists in a scrapbook of that time.

Lately when in Thailand, where we go for business and pleasure, to get away from Dhaka, I have taken to ordering morning glory greens along with curries. Dark green, lush, wet and oily, the oval plate arrives in a restaurant or in the small informal table and chairs on the beach where we are eating, in the dark, by a couple of can candles and the streetlights, not far away. We nibble on them along with rice and curry and whole fried fish, and I think of the varied uses of flowers, from a Somerville garden (where we sometimes sprinkled nasturtium heads on a summer salad) to a spot in Asia, how flowers accompany us and nourish us, our eyes and ourselves.

What is it about that color, morning glory blue, with its hint of purple and softness of value, that seems to be derived straight from the sky or from heaven? Like an Italian oil painting with its crisp clean sky, a Bellini Madonna, it's a color that speaks of coolness and brilliance. Transcendence, even.

Curry

The first curry I ever tasted was in England. I desperately wanted adventure after four years in upstate New York, and applied for graduate schools there. It was very cheap to live on the American dollar then. I discovered how very Asian London had become—particularly in its tastes. The straightforward English food contrasted with the Indian food. England: a pile of Brussel sprouts tumbling down from underneath a counter in a small vegetable store. A slice of crumbly Stilton cheese, firmly centered on a plate. Various kinds of pasties and shepherd's pies and crusted baked meat dishes, with leaves and decorations baked onto their tops. Ploughman's lunch, a hunk of French bread, a huge slice of cheddar cheese, and golden brown sweet pickle to sandwich between the two, served in a pub with a pint of beer or a shandy. India: an array of small dishes, all of brilliant color, with natural leaves and pods floating in them. A huge mound of rice. A pile of fragrant round breads, puffed and steaming.

Beer. The ubiquitous drink I had hardly tasted before coming to London, that flowed freely and warm from all the pub taps. Before the art school I was attending opened, and I found my crowd of artist friends, I stayed with Lisa, whom I had met in the States and her boyfriend, and they introduced me to beer and curry. One was expected to down a whole pint, if not more, on pub visits. Gradually the warm dark beer became familiar. My friends ate very late. One night they spent several hours cooking a potato curry, allowing the brown paste to heat and become golden as it rubbed off on potatoes and onions. Sitting down near midnight to a plate of curry and rice, I unsuspectingly took a large forkful. It was hot—very hot, and I drank long from the beer mug Lisa handed me. Burning—burning—then relief. It was a new taste, wholly unfamiliar to my New England tongue. I had managed to pick up New York tastes from my college friends, bagels and lox, coffee—but not international ones. Before long I was visiting the small Indian restaurants near Paddington Station, and generally feeling comfortable with the spread of exotically colorful bowls and delicious breads and rice that made up an Eastern meal. I came to like the lurid posters of Indian movies on the walls of their establishments and the loud colors of rugs and fabrics when I visited art school friends south of the river where many Pakistanis lived.

When I moved out on my own after a night or two with my friends, I stumbled upon the *Sani Gurugii* Youth Hostel, west of central London. There all we internationals shared a kitchen and I got to observe others' tastes, including the South African woman who cooked strong curries in order to warm herself up (it was late September) in this comparatively cold English air.

Unlike the curry powder bought in American spice cans, real curry is made up of a host of Asian spices, which people usually combine themselves. These include: cumin, coriander, red chilies, mustard, fenugreek, peppercorns, cloves and turmeric. Ground spices are toasted in a cast iron frying pan to the point of slight darkening of color and a delicate roasted smell.[1] When I first got involved with Michael I observed his individual bags and jars of spices, which he checked and replenished for freshness. Curry was no longer a small American spice can on the shelf, it was many containers, many tastes. I discover a new supermarket in *Banani,* in Bangladesh, near where I live, and musing my way down the aisles I stop at one large green tin of Madras Curry Powder from Bombay. Its list of ingredients are: coriander, turmeric, salt, chili, black pepper, fenugreek, garlic, cumin, bay leaf, ginger and gass, or Chinese cinnamon.

"Curry" comes from south India (Madras). *Kari*[2], or curry, is a fluid broth poured over rice. The variable spices of curry are mixed with coconut milk, also with vegetables cooked in *ghee* (butter) or oil: eggplant, onions, lentils, or as in the case of my first curry, potato. Spices are ground fresh, with a grinding stone, a stone version of a rolling pin on black granite.[3] Chili and cayenne were brought from South America in the 16th century, heating up curry's taste considerably.[4] In the beginning of the 17th century the exploration of the Americas also gave Asia pineapple, papaya.[5] Outside Asia the first curries appeared in Portugal, in the 17th century, then in England, in the 18th century, along with chutneys and ketchups and mulligatawny soup (pepper water in Tamil). Ketchup came to England from India via China; the word derived from Thai *kachiap.*[6]

The color of curry comes mostly from turmeric, a yellow-orange spice, and saffron, a brilliant yellow, which stains everything in its vicinity like a dye. My favorite Indian curry recipe requires saffron infused in a teacup with boiling water for ten minutes. During that time the dark red delicate traceries of saffron threads infuse the water with a deep, orange-red color. When I pour the saffron into the chicken curry, with yogurt, it forms a delicious deep yellow swirl with the white of the yogurt, until I mix everything thoroughly and tightly cover the pot, where it will combine with the other curry ingredients and the chicken and shallots, into a deep dark yellow. Saffron is the most expensive spice. It comes mostly from Spain now.[7] It smells something like tea, and is derived from the Saffron Crocus, locally known as *Karcom*. It grows in Greece, Asia and Spain. 1700 flowers yield 1 oz (25 g) of dried saffron spice, hence the relatively high cost.[8]

AN IMPROMPTU BASIC CURRY
Heat oil in a wok or a heavy iron skillet.
Puree onion till golden.
Add cloves, cardamom, cinnamon, then turmeric.
Soak saffron in 1/2 cup boiling water, for 15 minutes,.
Add ginger and salt.
Cook for 40 minutes.
Pour saffron water over yogurt, also add pepper and cilantro leaves.[9]

Abandoned Gardens

What is it about gardens, lost or uncovered or unweeded, that reminds us so much of another world or another time? They are reservoirs of lost color and memories and stories.

On the back of the farmhouse where my family used to go on summer vacations was an old stone terrace. This part of the house was more or less deserted, what had once been a kitchen, with a huge deep stone fireplace and built-in iron hooks for pots and deep shelves for baking bread. The shelves next to the fireplace held old children's board games; the bureau, drawers of old black and white photographs, of people whose identities were forgotten. The back door had an old wooden screen door attached and often the door would be open and the screen door in use, giving a glimpse of lush green as you walked by, and a whiff of hay, and leaves. To a child the house and land and gardens were all open to explore and ramble through. We got to know the slope of land of all the mowed parts, and guess about the presence of mounds and rocks in hills in the grassy, overgrown fields, as we traversed the narrow, mowed paths.

The stone terrace must have been laid nearly a hundred years ago. The step down from the back former kitchen, now a study or a quiet retreat, was steep. You held on to the wooden screen door for support when going outside. The stones were big-maybe a foot by two feet, and only the first half of the terrace was visible. The rest of it was overgrown with tall plants, some of them recognizable as having once constituted a garden. The terrace created a ledge jutting out from the house, which dropped off steeply at the end of the garden. An embankment held up the terrace and land. Running out over the lower lawn you passed its stony face, thickly laced with long grasses and capped with a border of old lilac bushes grown into trees.

I remember finding foxglove, creamy yellow-white cones on a long stalk, like a candle snifter or the fingertip of a glove; and peonies, sturdy old huge white-blossomed plants, pushing above the tangle. To a child most of the terrace was a mystery. Perhaps it harbored snakes—better not go too near. As an older child and adult, it was a place to reminisce about being a child, a state of anarchic freedom (a terrace is a world, a lawn and woods, a universe)—or a place to imagine sitting, among the sound of birds, the deepening shadows of afternoon, having tea or wine with a beloved friend.

The path from the door to the lawn bore to the right. The terrace was perhaps 15'x15'. At the end of the stone surface it changed abruptly to grass, a soft, well watered lawn around an old well, and the shadowy ditch which bordered the attached out buildings to the main part of the house, the current kitchen.

The terrace consisted of somewhat flat but far from perfect large stones with grass and weeds growing up between them like the lines of an organic grid. I remember once trying to play jacks on the surface of one (too uneven) and once drawing a picture with the side of a small stone, that left an imprecise, faint, whitish line. And in particular I remember the quality of sunlight in that place, which received light in the afternoon but which seemed to contain a kind of green, filtered inner light in the morning, when you pushed open the screen door and stood in the new air, filled with smells of wetness and budding plants.

The silence of that garden was amazing, as if it had sat for years, having once had a social existence, absorbing talk and laughter; then going to sleep, to commune only with the thin lines of nature that crossed it, birds, green stalks, and presumably, though we never visited then, winter. The land and farmhouse had been a convalescent home before my great aunt, who had a demanding job as head mistress of a school, bought it to spend her summers there. The house still had a variety of lounge chairs and chaise lounges from that earlier time, which we put to good use in the long summer days, lazily reading novels and old New Yorkers which were bundled all over the house in piles. As the oldest sibling, my aunt, my mother's sister, had inherited the house and land, and she invited us to use it in July, for our summer vacation.

When I lived in East Cambridge there was no back yard, just a covered back porch, looking out to someone else's backyard, overhanging the small cage-like space where Mrs. Papagna kept her angry dog. She also somehow had room to grow a single fig tree—and once she asked me to pick some figs if they were ripe. Not knowing, at that time, what fig ripeness consisted of or looked like, I plucked a couple off casually and handed them to her. When she touched them she looked at me with disdain. They're not ripe, she told me, and then I looked closely and saw the greenish-yellow quality of the skin which if I'd known about figs I could have easily told was young, unripe (something like myself, the term "green" feels appropriate). I felt, as she handed it back to me, the firmness of the fruit, not yet soft and pliable, like an older person's skin. My friend and I dragged old discarded black oil tins up to the top floor fire escape off my studio, filled them with soil and planted tomatoes. Italian plum tomatoes, appropriately, which we watered and eventually harvested, delighted at the possibility of gardening, even in the city.

Out that same window of my studio, facing the back inner city gardens, all pushed up against each other and guarded with posts, I could see a huge apple tree a couple of yards away. I watched it flower, and bud, and bear an amazing full crop of red apples, like a child's schematic drawing of an apple tree, joyous, resplendent, celebratory, in the midst of an old, anarchic city. It seemed to be growing there alone, without human aid or attention, except from those of us who unwittingly observed it.

In Ithaca I was offered a back space to create a garden by friends who lived in the country. I was all enthusiasm, but lacking in knowledge, as I accepted the challenge of making raised beds and planting vegetables. The spring that year was rainy, and cool, and I chose to bike all the way out of town to their place (as if adding challenge upon challenge in this task: a one speed bike). Several miles, many uphill. I borrowed tools from my friends, and after drawing up a quite ambitious plan, toiled away at the beds sculpturally, after having them described to me verbally, creating lower troughs to walk in. My energy began to falter after the first few trips, alone; I desultorily planted some cucumber seeds and abandoned the effort. Later that summer I went back to look at the intended garden and discovered what looked like an earthwork: the raised beds were still there, channels, walkways intact like an excavated city ruin. But nothing, but for a few weeds, was growing out of them.

At our current American house I reclaimed a tiny corner plot just outside the back door, which seemed to be providing nothing but stones and weeds. I forked through the soil, all 6 feet by 3 feet of it, throwing the stones in a corner, and added manure and compost. I planted herbs, simply, bringing small basil and coriander, mint, chives, thyme, and seasonally, lettuce and arugula. A manageable garden, a useful garden, one that I weeded, watered and harvested often. I learned something about expediency and proximity: the idea of a kitchen garden is to have it just steps from a kitchen! Not a long bike ride up hill away, not for my own use only…with no particular sculptural form except its own ever changing contour.

Borage

The first really successful garden I had was in Somerville. Just across the side street from my studio building where I also lived was a community garden, really just an abandoned lot which a zealous garden committee had claimed, surrounded by old brick and wood apartment buildings. Michael and I, in the beginnings of a relationship, asked for a plot, and together drew up a plan for what we would plant. I can't remember all the herbs and vegetables we planted, but I do remember wanting to try growing borage. It was the only herb I had heard of when living in London as a student. A friend of mine in art school had given me the hand written recipe for borage tea. Written after the name of the tea was "for courage." This tea was supposed to instill fortitude in the face of life's challenges, of which I had discovered there were many. I carried this slip of paper with the recipe on it around in various studio sketchbooks and notebooks for years. I had never thought to try dried borage. Now I had the chance to grow it fresh. It was also reported to have a very beautiful blue flower, approximately the color of morning glories. I was after another form of blue.

Before we knew it the borage had grown high and strong. Michael and I both watered and weeded the garden although I lived closer than he and got to enjoy it more. Then the borage began to take over. Its stems were 3 feet high before we knew it, its bright light blue blossoms everywhere, and it attracted bees. We were wary approaching this side of the garden. The garden had a kind of circular path, and we had set up a small bronze Buddha up in the midst of a cluster of herbs, and a flat rock to sit on, to make it a meditation spot as well. The borage infringed on this spot.

How would we control the borage? By late summer it was dense and climbing the fence on one side—and I had forgotten about making tea. We were trying to keep it in check to protect the existence of our other plants. There seemed, even, to be a war going on between the host of regular herbs and vegetables (we were enjoying tomatoes and cucumber and basil and parsley now) and this exotic experiment. The flame I had been carrying around all this while seemed to falter a bit—I had done my best with borage—with courage. Now another blue was in my life. I saw the energy of this plant and it was strong. Something to learn from, without letting it take over.

The next summer the city was making a tot lot in the space, and moving the gardens to the back perimeters. It was a noisy, hot summer. I had to keep my windows closed to keep out the dust and sound. It was a summer without a garden.

Mint

There are so many mints it is hard to have a personal relationship with all of them. Twenty-five species and many varieties make up the fan of mint experience. Spearmint is the most common mint, and most often infuses mint jellies and sauces.[1] Mint is so hardy that the same clusters of it are still growing at the old farmhouse I spent summers at as a child. One of the wonders of being there was discovering and re-discovering the plants at child levels. Just by the side door which we used all the time, with its large stone step and screened in porch, was a shapeless mass of bushes, surrounded by clumps of mint. A summer afternoon's pleasure came from slowly brewing up a pot of ice tea, clinking ice cubes in a tall old glass, running outside to pick a few leaves of mint, and pouring dark caramel-colored tea over ice, as the ice cubes shifted and settled, finally; poking a couple of leaves of mint into the top of the glass. The delicate flavor of the mint wafted into the tea, expressing the essence of summer. In subsequent summers no longer spent at the farmhouse, the inclusion of mint, which only took a little effort, a walk out into the garden, wherever I was, kaleidoscope-d me into the warmth and laziness of summers in the past. With the memory of mint, came the sap smell of August, as the trees heated up, cicadas droning, the sweet taste of the end of pink phlox flowers if you picked just one blossom off the masses that ringed the lawn, later the plop of horse chestnuts falling into the grass in the yard. Looking closely at Nature, smelling it, digging out the centers of sticks, mixing dirt with mud. Amateur primitives as children, some of us have grown up exploring materials and drawing inspirations from that same nature, longing for and revering the outside world. If it is possible to find a source of happiness it is this: living outside, being part of nature.

I found another use for mint in summer. Being devoted to my cat, I picked mint, chopped it up and added it to the baking powder biscuits I had learned to make in school. I had noticed that she lingered at the clump of mint to nibble delicately at it. Perhaps it was like catnip? I made button round small warm biscuits with green strands of mint throughout, and gave her one, brown and warm off the pan. Then I got the idea of putting real catnip into the batter, and made one magnificent catnip biscuit for the cat, who dragged it away immediately and spent a whole morning eating it and rolling over it, in ecstasy.

When we moved outside of Boston not long ago and I began to reinvigorate the herb garden just behind my house, my first purchase was a small pot of mint. I was pretty sure once I got it started it would return every year in abundance, and it has, along with the clump of chives. Summer mornings on my way to the studio upstairs in the barn I plucked a leaf of mint from the clump and walk slowly around the yard, checking out the tomato vines and cucumber, raspberry bushes and butterfly bush, all the while savoring mint's sharp, earthy, pungent smell. I'd take it upstairs to the studio after this ritual and find it days later, absently dried and curled on a surface. Recently I read that smelling mint increases concentration,[2] so I must have known what I was doing on some subconscious level. When I think of the end of the world as I know it I picture not only horseshoe crabs and cockroaches still marching around but huge clumps of mint, happily biding their time, thriving, until the next species discovers it and experiments with its use.

Kinds of Pepper

One of the basic ingredients of curry, staple of the East, is pepper. I never knew there were different kinds of pepper until I found a large tube of colored peppercorns in our local home furnishings and hardware store. This was a store on the edge of Cambridge, MA, and therefore catering to a more curious taste than the average home store, hardware store, or supermarket. But how had I managed to get through 40 years of life not knowing there were red, green and yellow peppercorns, not to mention white? I bought this long cylinder of peppercorns for my partner who is the natural cook of the family, because I knew he would enjoy trying them, and cooking with them. And, they were beautiful to look at.

The best-known peppers, black and white, are created, respectively, by drying pepper berries slowly in sunlight, soaking the berries so the outside husk rubs off, and then drying.[1] Pepper as an honest, direct spice of childhood—the one that sits ceremoniously next to the salt on American tables. Black pepper, connoting a little bit of daring in an otherwise safe and somewhat uninteresting American cuisine

Pepper has always been the most common spice to us. Even in a bland 50's childhood, there was pepper. The Classical world used it liberally, both as spice and medicine.[2]

It encourages appetite, and also is anti-bacterial. The fruit is used for rheumatism, headache, diarrhea and colic.[3] Since the second century AD pepper was grown in Kerala, India, and exported to Rome in exchange for gold. It was used as a preservative as well as a spice. Rome took so many spices from India such as pepper, in exchange for gold, that India had to turn to places in Southeast Asia to find a quantity of spice to satisfy the market.[4] Peppercorns and *tulsi,* or basil, were used locally in India as a cure for the common cold.[5] Pepper and cloves were at the top of the list of Columbus's prizes. Perhaps the original globalization impulse came from those 14th and 15th century explorers like Columbus who pushed onward into new worlds to find spice. His disappointment not to find himself in

Asia was mollified by his discovery of pepper in Haiti—a different family of pepper. He also found cinnamon and ginger there.[6] The East India Company, the most famous company of spice traders, was formed in response to a rise in pepper prices, coming from the Dutch and Portuguese trade.[7] Even the British takeover of India hinged on this response to spice monopoly.[8]

Michael and I collected peppershakers for years, and delighted in giving each other strange and wacky varieties: tacky, literal, sometimes plastic, as gifts. One of my favorites was a ceramic set that looked like smooth grey beach stone with a single line of white quartz delicately traversing them. One of the silliest, but most used, a chubby green ceramic cabbage from the 50's.

I was an enthusiastic pepper eater, loving that dust of black on fresh corn or summer squash, especially if freshly ground. But I became wary of its indiscriminate use when I read somewhere that it's one of those products that absorbs toxins most easily from the soil. I'll seek a source of organic pepper from now on, or forego it.

Green Tea

Green tea was barely part of my vocabulary till recently. I had tasted it in Japanese restaurants, and heard of the Tea Ceremony. But when I contracted cancer and needed to become expert on what to eat and drink to stay alive, I began to be familiar with it. Epicatachin, in green tea, is supposed to aid in the curing and elimination of cancer.[1] Green tea itself acts as an antioxidant, or an antibacterial agent; it also helps the liver to detoxify, and stimulates the immune system.[2]

Green tea is different from the other forms of tea in that it is heated in its entirety to prevent oxidation.[3] This tangential connection of teas to indigo is interesting—the oxidation process, and subsequent transformation of color or substance, common to both. Tannins are dry and astringent, which is how my mouth feels if I drink too much tea, and give tea its dark color.

When I had cancer, green tea came to me in many ways. A friend sent me a green tea candle, which I used in the middle of my table. Another friend living in Japan sent me a bag of green leaf tea, and a charm from a Japanese temple. Others came bearing boxes of tea bags, and hard candies. I tried the ice cream called Green Tea. I came to love the delicate clear simplicity of this drink. In brewing it I learned that fresh boiling water needs to be poured over the leaves and left for two minutes to steep, no more, to get the most out of its essence.

The three teas are green, black and oolong. Indian tea, a black tea which has a high tannin content, Chinese tea, also black, which has more caffeine, and which is supposed to clean toxins out of the body; and oolong, which lowers cholesterol.[4] All come from a small bush found in hilly regions in the tropics. In black tea the leaves of the tea plant are crushed, which exposes them to oxygen, darkening the leaves. For green tea, the leaves fresh picked from the plant are first steamed, then dried, so that no oxidation or fermentation takes place. The third tea, oolong, is only partly oxidized. Its medicinal properties fall somewhere between green tea's and black's.[5] From the Chinese trade with the Dutch in the 17th century, tea ran rampant in European taste. The first mention of tea is from 2700 BC, in China, and there is fossil evidence of ancient tea bushes. After 1657 the British colonies of India and Ceylon were planted with huge acres of tea, and China was no longer the sole proprietor.[6] In China it was believed that drinking tea promoted a long life.[7] In the 19th century Russia went crazy over tea. With the opening of many trade avenues with Asia, it was easy to import, and cheap to drink. The Russian process of making tea included a samovar; a teapot above a small stove keeping it warm.[8]

In the early 1800's, China tea was still imported to London. Varieties were faked; dried thorn leaves were colored with verdigris to make them black. About that time Indian tea was introduced, which had a different, warmer tonality. People found a way of faking this tea, too. Used tea was dried, hardened by glue and colored with black lead! In 1860 some controls were put on this and other subterfuges in the form of the First British Food and Drugs Act.[9]

Born of an Anglophile mother, in the home I grew up in, the teakettle was always on. The more British our domestic life, the better. My mother kept a cookie tin commemorating the coronation of Queen Elizabeth on the counter, although personally identifying with the younger Princess Rose. She was the younger of two sisters, so I suppose in her mind her older sister, who was being primed to be a missionary, an important task, was like Elizabeth, while she, quieter and more soft-spoken, was more like Rose. Coming upon a picture of the two royal children in an old newspaper in her attic I see that they were all probably about the same age, these sets of girls, so the comparison must have been obvious growing up, along with several thousand other sets of sisters. "Margaret Rose" hovered over my childhood like a delicate scent—the younger child, the less ambitious, the less pushed, yet in some ways, the more daring.

A black, enameled china pot with small bright flowers raised on its handle came out on special occasions, and a more mundane, dark brown china enamel china pot, as well as a Hooker's green single cup tea pot, were used everyday. When visiting at teatime, out come the teapot and the British tin. We each extract one round, plain sugar cookie, to dip in our cup of ordinary black tea.

When I went to college in NY State I began to drink coffee, enjoying Greek diners on Saturday mornings, but when I grew into my own way of being and life, I went back to tea. At first, herbal tea, as the 60's and 70's brought alternative cuisine to our kitchens and we reveled first in chamomile, then mint, and eventually endless varieties of lemon and cinnamon and fruit and spice. By now my friends and I find ourselves tired of these packaged herbal blends which we have drunk hundreds of times. Why is it that the ordinary black teabag, with milk, still soothes? I find myself experimenting with sage tea, and matte, other green varieties that seem to have an exotic and satisfying bent. And now in Bangladesh tea is King! In the north the tea estates raise a coarse black variety which we have become well used to, and like, and drink every day, interleaved with the requisite cup of green.

One weekend while in Bangladesh we took a train to the northeast tea-producing region, to *Sri Mongol,* leaving from the station in *Uttara,* on the outskirts of Dhaka. The old tea estates offer a private place to go for for a few days, one of the few places in Bangladesh with a guest house. After the noisy, crowded train, taking nearly seven hours, including delays; and hawkers and beggars roaming the aisles, people on platforms leaning into the windows, selling food and drink; the cool, hilly spaces came as something of a revelation. Quieter, more like a small town than Dhaka, activity in *Sri Mongol* centered around the train station and the markets which spread out along the main streets. A friend who was working in the region met us with a car and drove us to the tea station, through hilly country and winding roads, where you could see for miles because of the low lumpy vegetation.

Tea has always been grown in northern Bengal, now Bangladesh, and may have even supplied the earliest drinking of tea, in China. The British created plantations in the 1800's, and employed local Bengalis and imported workers, "coolies," in a feudal-like system, where the owners of plantations supported the tea workers, providing home, food, medical care, education in exchange for tea picking. The system resembles the plantation/slave owner system that was active in the American South, with workers treated severely for misconduct or insurrection.[10]

Staying in one of these old estates, a loosely gathered cluster of bungalows surrounded by miles and miles of low, green, bush-covered hills, was like stepping back into the past. Much of being in Bangladesh was like turning the clock back fifty years, but this tea estate felt truly frozen in time, deeply quiet, and apparently un-changed in many years. The close cropped grounds were manicured by a changing crew, and beyond the fence that separated the bungalows from the hills lay a small village of closely-arranged small houses, where the tea workers still lived. Beyond that was a low brick factory, where leaves were processed to become tea.

We followed a path past the village and over an ancient stile down to the road, and across it to the tea fields. Low stiff bushes lined the rolling hills, and would have created a sense of being in wild nature except that they were planted in such perfect rows. There was almost a surreal quality, a sense of being in a painting, in someone's conception of nature (which I suppose it was), with only the occasional brightly clothed figure of a woman picking tea.

The road itself was quiet, with an occasional rickshaw or truck or taxi. The following morning we got up early to see what was left of one the last original rainforests of Bangladesh, like visiting the last scrap of the Garden of Eden. Just two miles away the trees got very high on either side of the road. We parked the car and walked in on a dirt path, to a tangle of vegetation and trees. Light spun off the leaf canopy in streams and patterns. Parrots flew through vines. In the far distance we caught a glimpse of the furry, humped figures of gibbons in the trees, one of the last wild species left in what had once been a teeming jungle.

Returning to the silent compound we asked for a pot of tea and it came in a round china teapot. Aromatic and dark brown, steeping. With tea leaves floating on the top.

Red Crackers

Such a simple remark—and suddenly all the dyeing and the vegetable color and health concerns stack together like the children's wooden toy that is laced together with ribbons.

We are talking about our writing projects, and my friend Razia suddenly mentions some crackers her cook made for her. They were red, she says, and at first she thought no, she must have used too much red food coloring. But it turns out she had used *lalsag,* a kind of green akin to spinach, which has deep red/purple stalks and veins. When you cook the greens of *lalsag,* the water that is left is bright red. Her cook threw that in to the cracker dough, instead of plain water, and she ended up with red crackers. Similar to the color coming out of beets, which is a deep crimson red, and would also be a dye if added to a bread or cracker recipe.

What we need is little more color, I think, as I drive to *Aranya* one more time. I imagine biting into one of these bright red crackers and it brings to mind the colors one sees in India—in the food, the clothing, the decorations and painting. And to some extent in Bangladesh, although the Hindu community is so much smaller, and the influence less pervasive.

Maybe that's what tea does, flooding the body with color and its tannins/ polyphenols. Color, whether dye or paint or food, penetrating the very substance, the fiber of the painting, cloth, paper or body, flooding the cells, the interstices, the ground.

Today I am so versed in the names of these herbs and dyes I actually know what to ask for. Ruby and Moni are out, but they know I am coming, so I ask for indigo, *catechu,* and *kasmi,* but the workers say, substitute *hartaki,* another gray, instead of *kasmi.* I bring three small bowls of the dyes to the table. I am learning to come back to my first attempts, again and again, re-work them as I did in my studio, with other colors and overlays. Then they begin to look more like my paintings, which have layerings and scrapings and color coming through from underneath, a pentimento. I cover only 3 sheets of paper with marks, stopping early so they will dry enough to be transported, and while they are drying run up to the top floor where a new store selling crafts has a rack of handmade paper. I buy some more to replenish my supplies. As I run down the stairs with my roll of paper to see if today's drawing beginnings are dry, I pass the small quilt of dyed fabric on the wall which I admired in the beginning and a rack of beautiful dyed scarves, and enter the room with its long tables which is becoming like a second studio to me and I think: transforming material into color. The process of turning leaf, bark, flower—or story into deep, rich color.

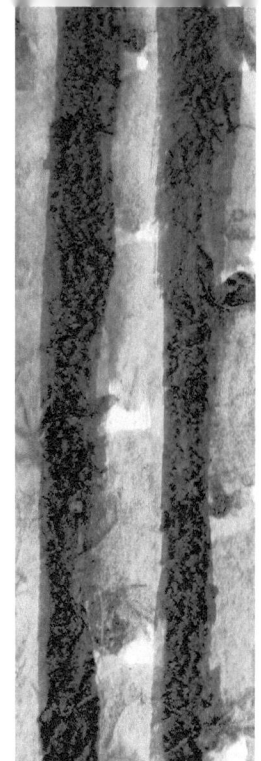

PART II

The Workshop

The project which is now the store and workshop called *Aranya* began in 1981. First called the Vegetable Dye Society, it was funded by the Bangladesh government small cottage industries corporation (BGSCI). Its goals were threefold: to identify and document dye producing plants of Bangladesh; to research and develop a whole range of dye-producing plants; and to promote usage of natural dyes, teaching weavers, printers and dyers to use them again.[1]

A main point of the project is and has always been to promote dyes that are eco-friendly. The Vegetable Dye Society studied the Indian example, and figured out how to adapt it to their needs in Bangladesh. Seeds were even brought to Bangladesh from India, and one of the chief technicians, Mr. Hussein, went to India to study its methods. The Bangladesh government Ministry of Industry gave the project a lump sum, and over two years they did research, eventually becoming a small independent business. From 1982 to now Ruby developed the project and the store to the point where it is known all over the world, with a range of dyes which exceeds India, which concentrates on printing, dyeing yarns and weaving saris.

Except for indigo, which had been a thriving product during the *Raj,* and has a checkered history here, dye materials in Bangladesh had been forgotten, except in tribal areas like *Rangamati*. Trainers coming over from India helped identify plants and teach techniques of dyeing and mordanting. They pointed out that most of the dyestuffs used are waste products, onion skins, marigold petals, leaves of various kinds. The most expensive dye materials, indigo and *manjit* (madder) come from the leaf and stem, respectively. Dyes are extracted from all parts of plants: leaf, petal, skin, root, bark, fruit, seed, sawdust. Indigo is starting to be grown now in Bangladesh, in an NGO or non-governmental organization called *Nikkomol,* "Blue Lotus," although large quantities of it in cake form are still bought from Indian sources. When *Aranya* was first started, Moni and Ruby found indigo plants growing wild by the side of the road in Bangladesh. When they crushed the leaves, after a few moments their fingers turned blue as the natural dye mixed with air. They brought back plants to

Dhaka to identify and experiment with. Indigo is the only blue from natural sources, besides the ancient plant woad, a cold-climate plant. *Manjit,* similar to indigo's unique blue, is the only true natural red, besides *cochin,* a dye derived from the cochineal beetle, used since ancient times.[2]

Recently *Aranya* studied the indigenous use of natural dyes in tribal areas of *Mymensingh, Sylhet, Chittagong* and Cox's Bazaar, where dyes are still used traditionally, and have not been totally supplanted by synthetics.[3] The *Koch* tribe, the *Garos,* the *Khasia, Mugh,* and *Mong* all use natural materials in coloring their fabrics.

I am just beginning a six day workshop on natural dyes at *Aranya* and I receive an e-mail telling me a friend's cancer has returned. I read between the lines that she is dying. This is after a summer of another friend's death, and last year, another's. I am learning of their dying, and I am learning about dyeing. This is not the first time life prepared me synchronistically for life and death, as I was picking up a new and important skill that would stay with me for a long time.

I notice that the bright colors are marigold, *manjit,* indigo (bright yellow, red, blue). But most of what is used at *Aranya* is subtle color—out of preference. Aniline dyes are much harsher in color than natural dyes, they have set a standard we are used to. In the workshop Moni says they mostly use alum and ferrous (iron) as mordants. These are easier, cheaper materials and they like the earthy colors that come out. They buy alum and ferrous in large quantities, and also use them for making printing pastes.

The dictionary offers the following definition of mordant: "a substance used in dyeing to fix the coloring matter, as a metallic compound that combines with the organic dye to form an insoluble colored component, or lake, in the fiber of the fabric."[4] Mordants are used only with natural dyes to make colors fast, something like the process of curing, preserving or treating materials. Normally natural color runs when fabric is rinsed. Moni tells me when she was young she and her friends used to dye clothes with turmeric, the dark yellow spice used so much in Asian cooking. It was easy to dye with it, and it made a lovely yellow, but as children they didn't know how to make it fast. Also in a ritual three days before weddings, the bride is covered with turmeric paste to brighten her complexion, and everyone's saris are dyed in turmeric to create a light bright yellow, although it would fade. Turmeric, related to ginger root,[5] has properties that heal. It is an anti-oxidant used against cancer, as an oil it works to relieve the pain of arthritis, and it protects against damage to the liver.[6] Takami, one of the workshop participants, says in Japan it is good medicine for the heart. I am finding that the plants and herbs I am learning about have multiple uses, that color can both adorn and heal.

Mordants can be natural salts or they can be chemicals. At *Aranya* four mordants are used, which are: Alum, Copper Sulphate, Potassium Dichromate, and Ferrous Sulphate, or Iron. An alternative to chemical mordants is using ash or sea water. And sugar, flour or honey can substitute for hydrosulphate, another ingredient in the recipe. In the past dyers relied on the substance of their pots to mix with the dye as it was boiling and act as mordant. Iron, copper and aluminum pots acted in this way, but the results were inconsistent and larger dyeing establishments found they couldn't count on the method. Each mordant elicits a different color from a dye. The dyer always uses the same percentage of mordant, the amount changing according to the weight of the material to be dyed; carefully measuring the mordants, balancing it with the weight. The measured mordant is dissolved in boiled water, and the fabric is immersed in the solution and boiled for twenty minutes. The solution is stirred at regular intervals. Cotton material is rinsed and squeezed first so the folds don't create ridges of uneven color. Instead of ferrous I learn you can use actual rust—let old nails or bits of iron sit in water for 3-4 days to dissolve. I will also use this to create ink.

In Colonial America cow's urine was added to the dyes as a mordant, when dyeing wool. When I told one group at *Aranya* this my interpreter and teacher, Moni, who also helps run the business, said maybe that's why the Indigo cakes smell so bad. She says the Bangladeshis themselves are too modest about bodily functions to experiment with using it.

I am understanding what I didn't at first: that just by mordanting the same fabric and dye differently, new colors result. The four chemical mordants look like this: Alum, a large white crystal; Copper, a deep dark blue crumbled matter; Potassium, a bright red/orange pasty crystalline substance; Ferrous, a light gray crystal. Easily accessible mordants also include ammonia, baking soda, cream of tartar, salt, urine, vinegar, washing soda, and less frequently, wood ashes, blood, lye, tree galls, clay, crab-apple juice, alder, sumac and aluminum foil.[7] Mordanting can be done before the dye is used, or at the same time the dyeing takes place.[8]

All this makes me think deeply about colors. I am slowing down to really see distinctions between these dyes. This results in very fine-tuned awareness of the essence of each color. Each one evokes a certain quality. I think of the old term "dyed in the wool", which never meant much to me before. Now I understand it means a certain at-onement between material and color, or a quality that can't be separated from itself, the object. I wonder how I can let experiences sink in more, take things slowly, appreciate nuances. I think about how experience itself colors life, and, finally, thinking back to that dream, how to soak the colors of food into my body to be nourished and healed on a continual basis, a kind of "cure."

Taking a dye I seem to like a lot I see that *hartaki,* when mordanted with alum, results in a light lemony cool yellow. With copper, it's a true beige. Potassium gives a light gold, and iron, a deep, warm gray. Eucalyptus, another natural substance I feel drawn to, gives a pale whitish gold with alum, a light brown gold with copper, a soft cream with potassium, and a medium gray with iron. I notice iron has the strongest color personality—dipping the cloth in it instantly darkens (technically called "saddening," the opposite of "blooming", or making bright) whatever color was there. Now as I see before my eyes the range of grays that are possible. Like beautiful dark lichens or bark in a forest on an overcast day, each emerges in its own right, with its own name and quality.

In *Aranya's* own book on dyes, *Rangeen,* the nine steps of dyeing cloth are listed as: "1. Weighing, 2. Calculating mineral and raw material, 3. Treatment of cloth or fiber in alum solution, 4. Preparation of mordant, 5. Mordanting, 6. Extraction of dye solution from raw materials, 7. Dyeing, 8. Developing, 9. Washing."[9] As usual when faced with many instructions I get confused and quickly lost about what to do. But fortunately the workshop is methodical and even repetitive. By the end of the six days I will have done the steps over and over again for different colors, and hopefully I will understand.

We meet in a bright back room of the shop, a screened in porch. The ceiling fan overhead keeps us somewhat cool, but the heat comes back when we're measuring materials and the fan is off. We are half inside, half outside, as we learn about these colors of nature and the ancient art of dyeing and rinsing, extracting and timing. Jamoli, a young Bengali woman dressed in a sari, is methodical and clear, giving us instructions step by step. I realize I am interested in making art materials from nature, not so much dyeing per se.

And learning more about color. I doubt that I will become a true dyer of cloth, but perhaps a maker of inks and pastes, which I will then use in my work. The whole process seems symbolic. It strikes me as the days go on that what the West's time-saving devices are doing (besides polluting the atmosphere and earth) are keeping us—for better or worse—from doing all the small measuring, stirring rituals which make up a relationship to earthen materials and actual measure. My young son's favorite activities have included sorting through beans, dyeing eggs, mixing batter, beating egg whites, making cakes; painting, molding clay, digging in the dirt, mixing plaster. He was learning so much about hand and eye and stuff, the stuff of the material natural world, as well as silently communing with me over the fluffy peaks of egg whites. My childhood was filled with these measuring and mixing experiences, which can be appreciated as creative and earthy, not domestic

Everything is done in slow and simple steps, no hurry. We go over everything again and again. I find myself thinking, having taught college for years and created a successful career, "this is how I need to learn." These feel like ancient rituals we are doing, none of which can be hurried or abbreviated.

I am the only one who is not fluent in *Bangla.* My Japanese friend translates what Jamoli says and from time to time Moni comes in to translate the longer instructions. I realize I'm on the other side now—not in the majority. I can use everyday greetings and get out some words in *Bangla,* but I haven't become fluent. Learning *Bangla* didn't come easily, when we first arrived.

Apple Trees

Learning a new technique like dyeing reminds me of other times when I've tried to pick up a new skill in a field that is not exactly my own, which leads to other things.

One year in school I decided to do as my science project, a study and practice of grafting. My father helped me buy and pot a small, dwarf apple tree. From my grandfather's back shed I borrowed an old cube of dark brown grafting wax, that smelled like resin, grainy and somewhat malleable. Studying the diagrams in the book I got out of the library, I carefully made a wedge cut on the small tree, and inserted a small branch I had cut at a diagonal as the book instructed. I softened the grafting wax and pushed it around the branch to make a collar. It also acted as a kind of glue, to keep the branch connected to the tree. I tried this in several spots on the tree, using different kinds of graft cuts and different apple tree branches, helping myself to small limbs from the town forest and a neighbor's yard. I wrote a paper, drew diagrams, and presented my project for the fair, dragging the heavy bucket that contained the tree all the way to the school auditorium.

The grafts did not take. Perhaps the grafting wax was too old, or I neglected some step which could not be communicated easily in a book. We planted the tree in the far end of the yard, behind a stand of bushes. The resilient trunk bore the notched scars of the failed science project; the configuration of branches found their own new pattern without the original or the grafted growth. It was a gnarly, small tree, which in my memory never bore any fruit.

My experience with the science project seemed to start a penchant in me for apple trees. The tree in Cambridge which I saw from my studio window, tucked into a neighbor's yard in a block of small old properties crammed together, came to represent a blossoming of self, of potential. Resplendent with branches, leaves and apples, I contemplated its bounty daily, and in every season. It flooded the collection of small yards behind tiny, run down houses with earthy cheer and a dazzling display of health and creative energy. In my awakening mood, in those years of artistic self -discovery, the tree was like a mirror of the best that nature and life could provide. As I moved out of that house, and Cambridge, to take up another chapter in my life, a neighbor chopped the apple tree down. I was shocked, at a loss to explain how someone could do this. But it was already time to move on.

I loved trees, and when it came time to buy our first house, Michael and I knew there had to be a big tree on the property. Alerted to the possibility of a two family for sale, we snuck over the night before we were to see the property, to check out the yard, the general location, the presence of trees. The trees on the street were very young, a new planting obviously replacing an earlier disappeared lane of old trees. Yet in the back of the house, at the very corner of the lot, was a very old, large maple tree, reaching its arms out over the small yard, which was bounded, providentially, by old garages on either side. Even in the dark of night we could tell this was a yard alive with nature and promise. The next day we saw the house, liked its spaces, signed a paper and moved towards ownership. We came to love that big tree that shaded the backyard, and sheltered our house, and our lives.

But I also had an idea of having an apple tree outside—it seemed to go with the idea of home, like a front door, a warm kitchen and a back stoop. I bought a young apple tree one Fall, choosing a red delicious type, and planted it on one side, next to one of the old concrete garages. For several years it failed to produce fruit. Finally one year, having read about the maturation process of fruit trees, I stopped the car on a road lined with apple trees, plucked off a branch loaded with apple blossoms, and brought it home. I danced around the apple tree, tossing the flowering branch against the blossoms in our tree, a latter day nature worshipper. For it turns out you need to have more than one apple tree to produce apples.

That year, the tree produced apples. Not big, gorgeous red ones, but individual, idiosyncratic greens. No matter—our apple tree had born fruit, and we were in the business of apples.

In Dhaka these past two years there were no apple trees, although we could buy apples in the market, said to be imported from Australia and New Zealand. But the need to live with trees, persisted. We could feel it in our souls.

In the first house we lived in, lined with apartment buildings, darkened by concrete, it was as if we couldn't breathe. There was only one jackfruit tree at the front, overshadowing the courtyard, its large rubbery green leaves expressive in the monsoons, plastering the pavement when the wind tossed them down. When we finally focused on moving, on finding another house, it was houses with trees that drew me. We ended up in our one story white house, picture windows opening to huge palm trees lining the road, their parallel-lined branches waving endlessly against each other, a kind of background music for the eyes and the soul.

Returning to our house north of Boston, we are surrounded by trees in adjoining yards, although not our own. Surveying the front garden, turned into a wild field in our absence, I discovered what appeared to be an apple tree that had planted itself at the corner of the house, thriving upward. Too close to the house to comfortably grow, it either had to be cut down or removed to another spot.

With care, one Saturday, I dug as deeply as I could around the base, cutting some roots inadvertently but saving as much as I could, tying a cloth around soil and stem. I pushed it in a wheelbarrow around the house to the other side, where I had chosen a spot, and dug a hole in the mid-back corner, just out from a line of yews. I filled the hole with water, and gently lowered the root ball in, tamping dirt around the edges. For the next few days and weeks I anxiously poured buckets of water from the dog's swimming pool onto the tree, hoping it would take. I wished I had a fish to throw in as I planted it, like Native Americans planting corn. Then we went away for a week, and when we came back many leaves had fallen off the tree. I trimmed some of the branches, hoping to stop the trend. In the end, the lower branches stayed green, and I kept watering it way into the Fall.

Spring will tell whether it has survived—and what kind of apple it is.

The Nine Steps of Dyeing

In the porch studio we work with squares of unbleached cotton, and fine rayon yarn something like embroidery thread. The tassels we dye look like wildly imaginative versions of soft corn silk, yellow and red, and beautiful off shades of green and gray. First, we wash the silk and cotton in soda ash, liquid soap, and caustic soda. The caustic soda is just for cotton—it is too alkaline or base, for silk, and would wear away the material. Washing the material removes the starch and other impurities so the dye will be effective. It's like the process I went through after cancer, cleaning out everything extraneous from my outer and inner life, to see clearly what the rest of my life would bring. A complete cleansing, to be cured.

In this process of dyeing, for 100 grams of cloth, we use 20 times that of water. We stir the fabric in the solution on the stove or hotplate for 25-30 minutes, using a stainless steel pot, for a consistent result. No matter how much fabric used, the time remains the same.

Glass and plastic receptacles are used to measure substances. These are piled up on the side of the room, along with glass rods for stirring. For mordanting, glass works best. The water is already boiling. To measure the soda and soap, we use a small old-fashioned set of scales, and a set of Chinese weights, cylindrical, with a knob for a handle, all nestled into their own leather box. I have seen these sold at the DIT 2 market near where we live, in antique stores that sell metal statues of Buddha and Shiva, and old terra cottas pieces of architectural ornament, wooden bedsteads and chests, Chinese coins and Afghani rugs. Our instructor puts a scrap of paper on the scale to keep it clean, and uses a small plastic spoon to scoop out the appropriate amount. She tells us, if you get used to it, you know how much is the right amount, a little too much is no problem.

We push the silk, which we are doing first, into the solution with a metal rod or wood chopstick, and move it around until it is wet, so every inch is washed. This is another activity where you have to take time, pushing and tending the cloth till it is all equally immersed. There is no way to make this go faster—I cannot hurry the process of absorption. I find myself, as usual, impatient with the process that exists, wanting to push it along. Slowly, slowly I poke the folded cloth in the solution with the implement, and see it darken as it gets completely wet. Washing silk is easier than cotton. It is softer, more slippery. Silk is a finer, more delicate cloth.

We used a large kitchen scale to weigh the fabric to be dyed. It is medium sized, bigger than a postage scale, and has a shelf with curved sides for resting the item to be weighed. It feels old fashioned, like something you would find in a kitchen in the 40's or 50's. After washing, the silk is rinsed in a pail of tepid water, rinsed in a second pail, then squeezed or rung out and hung up over a line outside. Then the water is discarded.

I can smell the hot pot on the stove as we are boiling the water. It smells of old-fashioned kitchens and laundries and soap. It brings back memories of various washing machines, even the hand-wringer type at the old farmhouse, hot soapy clothes, ironing, the crisp white flap of sheets on a line in sunlight, baskets of old wooden clothespins; the sound of an old washing machine spinning.

While waiting for some of our samples to dry, we visit the large blue indigo pot permanently placed in a corner of the courtyard, the "blue pot" referred to in so many indigo recipes, covered with its own roof. Everything around it is stained blue, the ground, the walls of the roof, the stirring implements. It is called the "mother vessel," and is never washed. At *Aranya* they never let go of the original dye bath. They keep dyeing with it until it is too light, then add more indigo and start again. Cloth can be left in for long periods of time to achieve a dark color or for re-dyeing. This concept, too, belongs to an earlier time—like yogurt cultures kept from batch to batch, or the activity of yeast, or the slow process of making wine or beer. A reminder that food, drink, or dye is alive—comes from living products, and has an existence of its own in time.

At Ossabaw, in Georgia, there was a saying that passed among us, especially at Genesis, the alternative art colony where we all worked in the garden, cooked, emptied buckets from the latrine, repaired buildings. "The island provides." It referred to the way we were living with the island and its resources. It referred to a greater wisdom than our own. The island provided: oysters, dug from tidal river beds; clean drinking water from a well; vegetables; old bits of pottery from the marsh, to paint; rare bird sightings; colorful snakes; "phosphorescence" at night in the waves; and dolphins, arching out of the water in pairs, accompanying our canoe. Since then Michael and I have borrowed this phrase when life seemed particularly miraculous: when we slowed down enough to notice gifts, whether from gardens, friends, or the dump.

In the workshop we learn that to achieve the best colors in dyeing, fabric should be treated with alum, one of the four mordants. Indigo can be used with alum first, or not. Alum, or aluminum potassium sulphate, is a water soluble crystal. Some dyers swear by treating all material with alum first. Alum is the least dangerous of the chemical mordants. You can eat it: the Japanese put it in eggplant pickle to retain the purple color, and to salve the throat. It's also used for filtering water in villages, and as a disinfectant.

White cotton cloth can also be bleached first, to whiten and soften the texture. The cotton will come out a pure white, affecting the quality of dye. Bleach is highly alkaline, like caustic soda, and is too harsh to be used on silk. We do a small cloth sample of bleached and unbleached cloth to see which will look best.

Wearing gloves, we mix the bleach, 2-4% by weight of cloth, into water. Making sure the cotton is uniformly wet, we leave it in for 10-15 minutes, stirring and moving the fabric all the time, so it is evenly treated, and using cool, not heated water. We rinse the bleach out in a clean bucket of water—all containers need to be rinsed carefully after use—squeeze out water, throw it out and start the rinsing process again, finally hanging it up double on the line. Again, my impulse is to get this all done quickly, so we can get on to the color, the main point of this morning. But I learn that all this preparation has a meaning, and if the cloth is not rinsed thoroughly the color will be affected by the bleach. I'm back in chemotherapy again in my mind, sensing the chemicals soaking through my body, a dangerous choice, yet meant inevitably to rinse out every cell that is possibly cancerous. How much patience it took to stay with it (was there any other choice?) when what I wanted to do was to get on with my life. But there might not be a life without that slow, slow process of destruction, then renewal. I was cloth, dipped, bleached and rinsed, and rinsed again, coming out clean.

Hartaki is another substance that can treat cloth. It is in itself a dye, used with black printing when done on top of red color. *Hartaki,* also known as the Myrabolan, a fruit, is used dried. The outer covering is discarded, the inner core used as dye. *Hartaki* seems especially to contain a transformational process, as the red dye will not take without the cloth being "fixed" first with it. A process of enabling, a transformation needed before future color can be attained. *Hartaki* itself is very good for stomach aches and can be found in herbal pharmacies as a medicine. Printing on fabric, the liquid is stamped, drawn or painted on top. The printing paste is more condensed, and contains glue. Some colors of inks are fermented (like iron and molasses, or *kasmi*), another indication of the living process of making color, and collaborating with nature on its product. To fix block printing requires a totally different technique, a developing system, like photography.

During the workshop the materials we use, dried herbs, bark, flowers, are clumped in five earthenware bowls in a round grid on the table, before us. They include: *manjit,* a soft chalky red dust, some of it in tiny stick form; *dalim* or pomegranate, a rind; *supari* or betelnut, ground brown seed; *latkan* or *annatto,* reddish chips of seed; *catechu,* dark vine wood; *daiphol,* light, diced red and green flowers; eucalyptus leaf, a dusty green, used locally for oils, incense, cold and sore throat drops; *shilkorai* or raintree, wood shavings and sawdust, from a big tree like an umbrella, with pink flowers. In some materials such as *shilkorai* or *catechu* you add a pinch of soda ash, also called sodium carbonate or washing soda, to make the dye solution come out faster. Each tree provides a different shade of color, as if expressing its exact nature, not just in terms of height or leaf or flower. It is even hard to get the exact shade again from the same tree. I like this difference.

Betelnut, also called *pan,* from the *Betel* nut palm, is an intoxicant common in Asia. The ripened fruit is hulled, cooked, sliced, and dried and placed on a vine leaf from the *Betel* Pepper tree, a relative to black pepper. The nut is believed to be an anti-carcinogen.[1] *Catechu,* also a cloth dye, is ground with lime and painted on the betel leaf underneath the betelnut—and chewed with tobacco. Traditionally *catechu* gives a dark reddish brown stain to the teeth in people who chew it as a mild intoxicant. Lime alone would burn the mouth—*catechu,* a base, neutralizes it.

To make marigold, or *genda* dye, we pull the petals out of orange-yellow marigolds and heap them on the scale. We have seen just these marigolds, these yellows and orange-golds, in strings of flowers at Hindu temples. Once when leaving Nepal the family of our taxicab driver, whom we got to know quite well over the course of our time in Kathmandu, invited us to their house. They served us fragrant soup and cokes from a nearby store. As in many Asian societies, it is an honor to have guests, no matter how humble the home. Asian families go out of their way to honor the traveler and the guest. We were given strings of marigolds and lumps of red paste on our foreheads, a form of blessing, and wished a safe journey. The hosts were as lucky as the guests in this exchange, one vaguely similar to the island/artist sharing I had experienced on Ossabaw. The giver is made fortunate in the giving, and a cycle of good will, health and partnership is initiated, benefiting all concerned. We put the petals in a pan of boiling water to cover, adding a pinch of soda ash, and leave for 30 minutes on the kerosene or electric burner. The flowers create a rich warm yellow orange color.

We pour the marigold dye through a wooden strainer, like a small papermaking tray or a screen over wooden stretchers, into a glass liter measure. More water goes into the residue in the pot, stirred around to get all the color. I squeeze the wet petals though the screen to get all the dye. Then we pour the dye into another large metal bowl.

We have rinsed cotton to get rid of folds and ridges. Cotton and unrinsed silk go into the dye bath together, and are stirred around. Although treated differently beforehand, when ready for coloration all materials go through the same process of dyeing. Through illness, thrown together with all kinds of people, I realized we were all having the same experience. It was enlightening to feel thus connected, even if brought about through painful circumstance.

The dye bath looks a cool lemon color, a light green/yellow. For 10-15 minutes we stir the dye bath over the stove.

Meanwhile in four separate stainless steel bowls we measure the correct amount of mordant, for 100 grams marigold petal. Each mordant is dissolved in boiled water, enough so the fabric is submerged well. We stir to dissolve the crystals with a glass or wood rod. Again, this takes patience. The mordant is done only when the crystals are entirely dissolved.

Fabric is squeezed of excess dye, then each of the 4 pieces is pushed into the separate mordant bowls for 10-15 minutes. Immediately the colors of the four mordants distinguish themselves on the silk and rayon yarn. Alum is a brilliant yellow, copper a golden yellow, potassium is a warm brown, and ferrous is a dark leaded gold. The miracle of these colors is hard to describe. Akin to the colors of leaves or the emergence of a photograph as it is developing—or the pure joy of painting, as a vision takes shape on a canvas.

The pieces of cotton come out slightly lighter when dyed than silk or rayon yarn. Cotton becomes a sun yellow, a medium reddish brown, a warm gold yellow, and a charcoal grey/green, with the four mordants, respectively. Each color is a unique variant of a classic color, with the bright yellow the closest to a pure primary color, and the others very much colors that might appear in nature, neutrals and gradations of familiar colors. Technically, most dyers mordant first, then dye, but Aranya has developed the opposite technique, to save and reuse the dye bath, which they used to throw away.

After mordanting, the cloth is rinsed twice, gently wrung out, then pinned to the line outside to dry in the air and sun. The colors look like late Fall leaves. The cotton is softer and washed out looking, the yarn and silk more brilliant.

Over and over I watch the white or off-white cloth absorbing dye and become something new and other, a clear, definitive color. Like becoming clear as a person.

In the last day of the dyeing workshop, we get to bring in something to dye ourselves. My other two companions have lengths of cotton and I have brought a bright off-white skirt, which I find too impractical for wearing. Sheleen ties her piece of fabric in an elaborate, long braid-like piece, and dips it, alternating, in *catechu* and *manjit,* creating a wild bursting red and brown tie-died sun form, very much like her personality. Takami has taken small toothpick-like sticks and wound them with fabric and string in small pinches all over her fabric. She uses indigo; and ends up with an asymmetrical design of round snowflake-like shapes, delicate and poised. I decide on a shade of *hartaki,* using both copper and ferrous to get an in-between taupe color. I measure the fabric and the *hartaki,* boil the water, and put in the *hartaki* rind and a pinch of soda ash to facilitate the dyeing process. It is so hot in the room we are dyeing in—an overhead fan, an occasional small breeze, but mostly this humid, oppressive heat where you feel a thin layer of sweat on your skin at all times. You notice it, then you forget about it, the weather is like this so much—and then you notice it again.

I've stirred the dye and will leave it for 30 minutes. I strain the *hartaki* out, swirl the bowl with water and add the remains to the strained dye. I pour the dye back into a pot, rinse the skirt in water and add it to the dye bath for 15 minutes. After stirring it in the copper sulfate mordant for 10-15 minutes, it comes out a dull beige. I hang it to dry and ponder this color. I'm disappointed, the color is so ordinary. But then I experiment with a small amount of ferrous in the copper, with the help of Mr. Hussein, a ¼ of 1% solution that gives just what I want—a taupe color, a grey-brown earthy cast to the skirt, like wood or mushroom color, full, rich, and utterly, completely unique.

We spend the last part of the workshop threading samples of yarn we have dyed onto cards and labeling them. There are distinct differences between the way cotton takes color, and how the rayon yarn does. Here is nature expressing itself again, I think. Each material interacting with the substances of the earth in a particular way. A kind of infinity of response, like the awareness that dawns on you as a painter, as you discover more and more color in what had appeared to be one color, gray or a brown. The revelation of looking around the contour of an object and finding slight shifts of color as you go around. How light affects the perception of color, and atmosphere and other colors.

Manjit cloth is a bright rust red, and *manjit* yarn a brilliant deep red. We take home both yarn and cotton samples, stapled together in squares. I marvel at all the work that went into them and how these colors have come directly from nature.

As the class winds down we are all quiet, writing in our notebooks. Sheleen and Takami, who work in *Rashahid* in the North for a youth employment program, will be implementing a natural dyes program for their students. I plan to learn to make ink and pastes that I can use in my studio. I am thinking about the process of dyeing, of deeply infusing fabric—or anything with color that comes from nature. This changing of a thing, like a transformation or process, like Fall, or aging. I learn that the colors in dyes differ depending on the time of year, and the softness and hardness of water.[2] That the deepest colors come from simmering the dye at low heat. They get dull and brown if overcooked.[3] That you can tell what kind of color the plant will yield by rubbing it against paper[4], and that plants need to be gathered when they are young and vigorous, roots collected in autumn, berries and seeds when ripe. Lichens are gathered in August, when they are most acidic and the colors strongest. And after a rain, they come off most easily.[5]

All this feels like a poem to me, a learning directly from nature.

Seeds

In Dhaka as the natural color paintings develop, and as I gain momentum, skill with the brush and integration with my new medium, the marks I make sometimes begin turning into shapes I can almost recognize. Particularly, the big brushy strokes resemble seeds, and in a flight of curiosity and enthusiasm I go with the idea, pushing the strokes a bit more into shapes and using the wrong end of the brush to scrape out a side or define an edge, or describe the vertical lines scarring a seed's body which help it plant it more firmly in the earth.

The seeds seem to be raining down, streaming, in a torrent of energy intent on the earth. At other times the seeds turn back into marks, big bold splashy marks, or delicate smaller ones, still falling with abandon, heading for the ground. *Acorn Earth* is related to the seed paintings which came spontaneously from using natural dye. I had to work and re-work the shapes and shadows to get the dark tones and rich browns of the finished piece. *Acorn Earth* is like a homage to autumn and the continual renewal that is everywhere in the growth processes of Nature.

Marks and Water uses blue in long descending strokes juxtaposed with dark ribbons of brown black, and lighter iron and molasses underneath, to form a sheet of descending water and strokes. Other, larger marks, as in *Marks I-IV* lay large, reddish-brown brushstrokes of natural color over smaller indigo marks to create stream of abstract energies. In *Brush Painting I-III* the strokes become so large they almost obscure the colors underneath, becoming a continuous rain of color that is no longer individualized.

These marks and seeds relate to the black dots and dashes of my ink drawings, that came right out of nature, a language system of marks to stand for the intricate web, as I saw it out my studio at the MacDowell Colony in the woods in New Hampshire. Stumbling outside with ink and a pen, with bits of rough watercolor paper left over from a painting, I attempted to put down without actually drawing it, the lights and darks outside my cabin studio. Where the paintings so far had attempted to describe nature, I was trying to take out the intermediary design and concept and be wholly present to its essence.

Now the ink drawings, which I have been doing for twenty years, and the brush drawings using natural colors, are coming together.

Painting Notes

Old batik work can be worn down until the fibres

begin to break, but will still retain the indigo-blue

with almost its original depth of tone or luster. [1]

In a dream I had in California in the early 1980's, an American Indian woman accosted me, etching a symbol on my palm, which meant "of the earth." The symbol was like a primitive mountain shape, a triangle with an open side. The paintings I did around that time, when my work was about to metamorphose into abstraction, were landscapes seen through a triangle. Like the fan shape but up side down, the triangle captured space, but no longer hid it. Openings, usually to watery fields of marks and texture and light, they used landscapes I had seen in my first trip to Asia, in 1985. Sometimes figures were superimposed on the landscape, a meditation on human scale and connectedness, or disconnectedness. The landscapes were brilliantly colored, the figures usually black and white, a combination I couldn't explain at the time. In about eleven months I painted thirteen triangles, large and small—and then my work merged with the water.

Peeling back the layers of figuration I had worked on over the years, I was going for something deeper, perhaps what nature wanted to say, through me. In a way, looking back, the triangles were about a loss of a sense of being at home in the world. In one, *Out of the Garden,* figures borrowed from a Renaissance painting, weeping, leave a breathtaking red and yellow visionary landscape. In another, *Creation,* a woman walks out of a landscape, at ease, like the earth, but still in black and white. In yet another, one of Michelangelo's "dying slaves" is juxtaposed over a tropical paradise reminiscent of a primitive's painting of jungle. The earth just is; human beings, tormented, try to wrest themselves out of a clutch of identity they seem not to even understand.

Now in my meditation on blue, on indigo, on colors from the earth, I learned about where color came from and also about how to be at home with the earth, be gentler with myself in relation to art and its production; how to enter into a collaboration with "what nature provides." Letting go of my own imposed ideas opened me to the magic that comes through process. Materials assert their qualities. Trusting that the experience would take me somewhere meaningful—the archetypal quest similar in art and adventure—I put down my oil paints and my heroic scale; my push towards super productivity, and entered into relationship, listening for what nature, color, the earth had to tell me; entered into a reciprocal involvement where I would use what is no longer needed or is unheeded, to make something else: ink, color, visual field.

★

I get down on the floor and spread out the banana and water hyacinth paper, made locally, using local materials, and what is left of the dyes from *Aranya;* with big and small brushes cover the paper with marks. In doing this I realized that connecting with landscape as I begin, focusing on it even in photographs, is what gives the work some of its power. Then the landscape is with me as I move the brush, as I dip into the dye, as I stroke across the paper. The landscape is the ink and brush and the paper.

It feels as if I've been living with this work for a long time. The paper feels right, the ink feels right. I'm home again in my work, in my body, although not technically "home." The studio, finally, is warmed up.

★

Nature, people, and color. The history behind the production of colors. The relationship to land, to "home." While I've missed New England—its weather and landscape, its familiarity and hilly terrain, I seem also drawn to tropical locations. In the luscious, flagrant growth of the vegetation my artwork seems to flourish. *Palm Marks,* a small 18"x24" canvas I did in my first year in Bangladesh, is a riot of lines of palm branches I looked at every day from my window in Dhaka. It's not a literal view of palms, but a visual collage of impressions of those long, sweeping lines that etch the sky when you gaze at a tangle of palm trees. In the center is a doorway shape, also connected to the top of the canvas by a line. In the doorway, the lines are slightly larger scale, the palm stroke somehow softer, as if looking within afforded a slightly different perspective. Like a view of history, perhaps, or the change of awareness that comes from living within another culture.

Brilliant blue intersects the palm marks on the left, a reflection of the white wall behind them in our yard, which was continually changing colors depending on the time of day or the shadows cast. The blue offers another kind of opening to the eye—an open space, or place of rest. No one can bear to look at total activity or unending excess of line and mark and pattern—the eye shifts to a place where there is no pattern, one single color, or uninterrupted space. I learned this in my first landscapes, as I painted in the backyard. And it became part of my understanding of filling space.

In *Palm Marks* the blue is of indeterminate space, but holds the eye down for a moment before it starts tracing the lines across and up and down in endless movement. In the blue, there is substance and richness, and a kind of stasis. Shapes are defined by edges of palm, but the shapes, unlike the long strands of color defined by palm branch, are closed. One's eye would stay there anyway, for a moment, in the warm deep satisfying color of blue.

Planting Blue Gold

Digging deeper into the story of indigo production, I find that when Eliza Lucas Pinckney, the woman who so successfully created an indigo business in America, left the islands to supervise the S. Carolina plantations, an indigo expert from the French Indies came to help her produce indigo dye cakes.[1] Interesting how the abbreviated story gave her all the credit, still a relatively young girl, for managing the indigo harvest herself. "Experts" and "helpers," undoubtedly all of color, make the indigo revival happen.

Earlier, indigo, a cold-water dye, at times called "blue gold," was brought from the Indies to sell on the street in America for $2. per blue cake. Woad, the other blue dyeing plant, which smelled very bad when fermenting, which indigo replaced, used up the nutrients in the soil, requiring rotational planting.[2] Louisiana especially relied on indigo production in its economy, and the plant now grows wild throughout the state.[3] In Mexico, *anil,* or indigo, is still cultivated and made into cakes.[4] Many plants come from one indigo family: India and China share one branch of the tree, as do South and Central America. The East Indies originated another, that spread to most of the tropical world. West Africa hosts another branch, as does Japan.[5]

I really begin to understand the indigo dyeing process when I take a trip with the Asia Study Group to *Muktagacha,* a *Garo* community west of Mymensingh, north of Dhaka. The *Garos* are one of the many tribal peoples that live in Bangladesh. Originally from Mongolia, discriminated against in the wider Bengali society, these tribes were pushed into the northern and hilly areas of the country in the 13th century.[6] I had missed the trip to the *"Neel Baris"* or blue houses in *Mirapur* earlier in the year, big palaces owned by wealthy British merchant families where indigo was brought by boat down the Ganges from Bengal. There is a bad feeling, I learned from Moni, still, in Bangladesh, among old farmers, about indigo production. They were reluctant, even afraid, to start it up again. The association is with killing and oppression, even famine during the Indigo Wars. But here I get to take a look first hand at the plant, the small industry in its second year, and the process of dye extraction. The Mennonite project, which is now focusing on indigo, tries to extend crop patterns and create sources of revenue for people in the countryside.[7]

We gather at 6:45 a.m., on a Saturday, at Virginia's house, which is not far from where I live. Twenty-five of us share a rented bus from the only tour company in Bangladesh, called "The Guide." The way is slow, taking us three hours to get to *Mymensingh,* where we stop to meet the directors of the project, and see an actual indigo plant growing. In the courtyard of a small building that houses the offices of MCC, the plant looks like a feathery bush, medium green and light. It has tiny leaves and miniscule pink flowers like bleeding hearts. From the branches hang one- inch long thin seedpods, like miniature green snakes. Indigo is in the legume family, like beans, and won't prosper in a pot. The plant needs well-drained soil. Land that is partly hilly will do. The leaves are cut the first harvest, the stalks the second, and the third time the farmers let the whole plant go to seed. Like legumes, the indigo plant gives nitrogen back to the soil. The low full light bush is easy to harvest, cows and goats don't eat it, and it has no problems with insects or viruses. It needs no insecticides or fertilizer. Thus it is very promising for farmers. One acre of indigo yields 3 tons of green indigo plant, with everything harvested. Lately farmers in the area have been planting pineapple with indigo around it as a natural fertilizer. Most impressively, one field yields fifteen kg of indigo dye. Costing 3500Tk (roughly $61.) to produce, it yields 15,000.Tk ($263.) in the marketplace.[8]

We travel another half hour to the *Modipur* Forest area. The sun feels clean and bright. There is so much less traffic on the single lane roads, the air is relatively clean. We drive through intervals of field and forest. Finally we stop at a small house with a walled-in front garden growing cabbage and squash. Walking through the front room and out into the back compound, we are suddenly in the midst of an indigo industry. On the packed dirt floor of this perhaps 20x20 foot space, are the various stages of indigo dye production. On the right are large plastic blue and black drums where water and indigo pass, and pumps and plastic hose carrying indigo in its various stages to the next station. Treadle pumps move the water from one container to the next, as well as gravity, with some barrels poised higher than others.

The courtyard smells like wet hay. Huge bunches of green indigo stalk and leaf, in sheaves, are submerged in water for twenty hours, with bricks weighing them down. The pigment of indigo is in the leaves, but the whole plant is used, and then the leaves can be saved and used as mulch, or sold as firewood. After twenty hours draino-like white flakes of caustic soda (sodium hydroxide) are added to the water, while it is beaten with wooden beaters made of bamboo, one end split and made into a round cage. *Garo* women have the job of beating, which helps ferment or oxidize the dye. The sun heats the indigo and water, which turns green, from the plant's sap. Air bubbles develop in plant roughage and come to the surface. Greenish blue foam appears on the top of the vat of constantly moving water. Caustic soda helps to settle the sediment. The water is let out and the sediment, only a couple of inches, which looks like indigo mud, is saved. The indigo is further refined by sieving through a cloth into a large bowl, then dried into flakes that look like thick blue rose petals. These are ground, and made into a powder, which is then sold commercially. The grinding is done with a stone roller, like ones used to grind spices. Where powdered cakes are made, the indigo ferments in a water solution of ashes, cow's urine, acid from beer, yeast or rhubarb juice, in clay pots, according to local tradition.[9] India sells its indigo in cake form, a very dark lapis-like blue. The dye is prepared at least three days before using it.

The goal is to get Bengal Indigo on the wholesale market. Six hundred acres of land are available and possible for indigo cultivation, though right now only 12 are being used. So far only three hundred kg. of indigo have been produced, while the market demand is 300 tons. One ton of green material yields 5kg of dye. One tub produces about 400 grams of indigo. Pure indigo has nothing added in its processing. India adds a bark solution, Thailand adds lime instead of caustic soda, which gives more quantity, possibly due to the lime weight, but the color is less strong, the blue whitish. Fortunately caustic soda changes in the process of making indigo dye, becoming almost neutral when it is poured back into the ditch as waste.[10]

We wander around the courtyard, taking photographs and asking questions. There is a constant sound of running water and talk, and the damp, fermenting smell of the indigo branches. We are like flock of birds restlessly moving about, then settling for a moment, then moving on. Small cups of strong dark Bangladeshi tea with sugar are served, on a tray, some in small glasses, as is the custom in the East. We linger some more, then head back into the house. In the darkened front room we gather to hear more about this particular project, how it began by experimenting with a few plants at the front of the house, perhaps the one of the ones we saw, and how it will take 3-4 years of experimenting to increase the yield. The potential is great for families to sustain themselves by growing indigo. Dye will be a profitable export.

At one point one of our guides refers to the history of indigo being tied up with British imperialism, and there are several discreet clearings of throat as the British members of the group mutter something about colonial experiments, gently protesting. Michael and I look at each other sideways, smiling.

So far *Aranya, Kumindini,* another craft and clothing venture in Dhaka, and a few other customers in Bangladesh buy the indigo. "Source", the MCC store downtown in Dhaka, also supplies it. Paper products using indigo are found in "10,000 Villages" and Oxfam shops in the US and UK.

Why was indigo such an important dye, for so long? Why was it called "blue gold"? Available in many parts of the world and in use for over 4,000 years, it is readily used with animal and vegetable fibers, and it is extremely color-fast.[11] The leaves have other uses: they are used as a black hair dye, and in China the roots and leaves are used medicinally for depression, swollen glands, heat rash and cancer.[12]

But I'm tempted to say, people need the color blue.

The Mirror of Indigo

The entire indigo industry of Lower Bengal ultimately

rested on a foundation of coercion and intimidation...[1]

In doing background research about indigo, I discovered the existence of a play about indigo produced in 1860 in East Bengal. It was shockingly explicit in its depiction of violence toward Bengalis, which "result[ed] in either the madness or death of almost every one of the principal Indian [Bengali] characters in the drama".[2] The play lashed out publicly, cataloguing for the first time the strife endemic to indigo production leading to the indigo uprisings. Called *Nil Darpan,* "The Mirror of Indigo," it dramatically expressed Bengali feelings about the indigo planters and the conflict that created the indigo wars in 1859-62. British authorities were confronted with the play by a socially concerned missionary, James Long, in 1860. Its repercussions eventually led to a complete change of policy towards indigo workers.[3]

The play was written by Dinabandhu Mitra, a post office inspector and playwright.[4] In the play all the abuses surrounding indigo cultivation are compressed into one story. Peasants cultivate indigo for virtually no money; planters corrupt and abuse local Bengalis and British subordinates; women are forced into prostitution; planters engage in a culture of extreme violence.[5]

The play was translated, published, and distributed in Dhaka—but also sent to Indian and London newspapers, members of Parliament, and well-known officials and corporate players, without the Lt. Governor of Bengal's knowledge.[6] A huge scandal ensued, and James Long was put to trial for libel in 1861.[7] He was accused of slandering the editors of the pro planter newspapers, *The Englishman* and *Hurkarn,* and the planters as a whole.[8] He was found guilty, and sentenced to a fine and one month in jail.[9] A wealthy Indian patron paid the fine, and Long was visited and addressed by huge numbers of officials, Indians, missionaries and leaders of the Hindu community, who supported his effort.[10] The result of Long's imprisonment and his stand on behalf of the peasants was that the upper class of educated Bengalis became conscious of the rural poor and their issues entered into the politics of the time.[11]

The main conflict in growing indigo was Bengali farmers' preference to plant more profitable crops of rice or other food, instead of indigo. British concerns led the push to develop indigo as a commodity. The fickle weather of Bengal, with its unexpected storms and floods, further made indigo cultivation difficult.[12] As well, Indigo workers were treated outrageously. "Coolies," the lowest paid, lowest class workers, often tribals from northern Bengal, climbed into the dye vats during the process of indigo production, walking within the vat, continually beating the water with paddles, effectively oxidizing the indigo with their bodies.[13] Today the stirring motion is accomplished from outside the vat with long sticks or beaters, or achieved mechanically.

A shifting web of economic and social relationships supported the production and export of indigo. In the early days of indigo cultivation, the British Government of Bengal gave monetary incentives to planters to cultivate it. Originally planters had to pay peasants to grow indigo on the land they leased from Bengali *zamindars,* or local landowners. Later, planters could hold and control the land independently, although still acting as landlords under the *zamindars.* Planters and *zamindars* fought out their disagreements using locally hired bands of warriors, called *lathiyals.* Planters could hold a variety of kinds of leases, but tacit assent was needed by the *zamindars* before the planters could make the peasants cultivate any crop. Banks and agency houses in Calcutta eventually regulated the flow of money into indigo. Over the years many of these went bankrupt, throwing the industry and the price of indigo into new chaos. Fights took place between planters, and there was corruption in the courts even when disputes were brought before magistrates. British colonial interests saw indigo as a way of taking their money back out of India and Bengal. In the "remittance system," the original investor in indigo, the British East India Company, was able to recover costs of loans to Indian traders, through exporting indigo from Bengal, where it was sold on the market in Europe.[14]

In 1858, in one of the frequent shifts in administration in the region, independent white grower/investor planters were made honorary magistrates in their regions, while the Indian *zamidars,* or local lords, had their political power taken away. Much protest followed this decision.[15] The then current Lt. Governor of Bengal was fearful of loss of control following the racially motivated Sepoy Mutiny. Anger at the loss of local control, plus civil servants' refusal to acknowledge the planters' powers, incited what became to be known as the "Indigo Wars."[16]

Earlier, the Farazi Disturbances, organized by a Muslim group with political ties to Calcutta, had resulted in the destruction of indigo factories in 1832. By 1859 the disturbances directly organized by Farazis had been stopped, but many peasants participating in the Indigo wars were Farazis, skilled in weaponry and fighting.[17] Other revolts included the Santal Rebellion of 1855-57, where tribals revolted against control by planters, *zamindars* and Hindu moneylenders; and the Sepoy Mutiny, in northern India in 1857-58, a revolt of native troops which instigated the transfer of administration from the East India Company to the British monarchy. The 1858-59 "wars" emerged from earlier revolts, while adding the element of a well articulated and new sense of social justice. The revolt stemmed from a skewed system where local people were judged by a different system than the British who lived in Bengal, much like the slave system of the American South. Bengalis knew they were being forced to plant indigo, not for their economic gain, but for the British planters' gains. Eventually, more enlightened voices, including that of John Peter Grant, the Lt. Governor of Bengal, recognized and spoke out against abuse heaped on native workers. He concluded in 1859, "Indigo cannot be supported at the expense of justice."[18]

During the indigo wars local fighters were hired by planters and *zamindars* to protect their interests, and by 1859 whole villages went to all-out war with hand-made weapons: clay pots, brass plates, slingshots, bows and arrows, and spears, to fight the forced growing of indigo.[19]

In 1861 magistrates were advised not to force indigo production on the peasant farmers. Farmers subtly undermined the indigo factories by refusing to sell to or serve the managers of the indigo factories in any way, forcing them to leave their jobs.[20] In 1860-61, indigo workers availed themselves of recent changes in the judiciary in Bengal. Grant had multiplied the number of courts, and made them available to the peasantry, shifting power from the executive to the judicial branch of government and administration.[21] By 1861 peasants were filing lawsuits against planters, having developed a consciousness of their rights and the legal process. Previously problems had been resolved by edict and authoritarian rule. Now peasants were seen as equal participants in the system.[22] For about ten years this system favored the lower class, but eventually the rural middle class rose into a position of prominence and power. Many in the middle class were moneylenders. Earlier they had backed the indigo workers in order to bring about the downfall of the planters. Now the moneylenders aligned themselves with the legal system, which supported their activity. Local law, honoring the rights of the poor, dissolved.

By 1862-3 a new calm descended in Bengal. Some peasants went back to cultivating indigo, and were paid twice the amount they had been before. They did not have to use their best land for indigo, but could plant other crops at will. Many planters abandoned indigo plantations and got involved in other work.[23] The basis of the indigo wars: race, class and economic justice, helped set the stage for the future independence of India,[24] and, most likely, later, Bangladesh.

Bengal today is self-sufficient, but it struggles with the need to develop more industry, and increase its revenue through exports. The shrimp industry, the second largest foreign exchange earner, wrestles with issues of quality, environmental concerns, and foreign markets' lack of trust in buying products from this small South Asian country.

The re-emergence of indigo as an ecologically friendly business benefiting the small landowner and businessman in Bangladesh is an amazing reversal of its earlier status. No longer linked to colonialism, big business and personal fortune, individuals and small non-profits and businesses have wrested it from a bloody history tied up with slavery and racism. The Mennonite revival, *"Nikomol,"* and *Aranya's* brisk business in cloth and clothing derived from Indigo and other plants, represent a new start in conscious husbandry and economic gain.

When I encountered indigo for the first time at *Aranya,* I had no idea how much its history reflected the history of Bengal: the color blue, woven like a strand into the tapestry of South Asia's struggles and independence.

So deep it is part of the actual fabric.

Blue

A piece of cloth dyed in indigo can be safely washed

and even boiled with white cloth without affecting it.[1]

The most poetic of the twenty one entries on blue: number ten, "the blue: a. the sky b. the sea c. the unknown" and number nine, "out of the blue, suddenly and unexpectedly." Perhaps blue is about limits and the experience of limitlessness. It is about depth, and stillness and vastness. Gazing at a lapis pendant from India bought at an outdoor fair in Dhaka, I am drawn to its glints of gold and silver, its irregularity. The blue a deep medium blue, seeming to go on and into its dark blue space without end. Lapis lazuli ("stone like the azure sky") was ground for pigments in the early history of painting, and is also the color of royalty in many cultures.

Indigo with its history of richness and mystery, tainted by its association with slavery and ruthless mercantilism, seems to have an indelible nature in the world history of dye. It is one of the few dyes that does not require a mordant, or fixing agent. Hence the quote above, from the recipe on p. 31–32 for indigo dye. But perhaps it is also the fact that indigo is blue, that makes it a lasting presence. And the fact that it is one of two blues from natural substance, the other being woad.

 The feeling that blue was here before we were—and that it will be here when we depart—imparts a sense of the ancient and eternal to this color. It is not going away, one feels. Here to reflect our own creations, our complicated or simple life.

Synthetic indigo completely usurped the natural indigo trade by the early 1900's. Constant in quality and color tone, its industry could be maintained in any climactic conditions.[3] Blue became a product from an industry instead of a yield from the earth. Between the Indigo Revolt and the hybridization of the process, indigo was almost totally lost to many locales such as Bengal.

What happened to blue? Blue aniline dye was imported to Asia—including Bengal—instead of Asia exporting indigo in great quantity. Synthetic blue dye being far cheaper, blue clothing became easily accessible in a way that it had only been locally. Blue jeans, representative of democratic and globalized culture, are dyed with synthetic indigo.

Blue—memories of warm mimeographed pages with a familiar chemical smell, right out of the machine. Brilliant blue popsicles melting in the summer sun. Blue berries dark and dusty, not to be mistaken with dog berries: fat and succulent on a forest bush. Blue prints rolled up with their grainy blue lines, a cryptic language of spaces and measures. Blue plumbing chalk, for lines and measure. Dusty blue strings. Blue stones under water in a stream, drying in the sun, to gray. Blue-gray dogs, whippets and greyhounds. Blue Egyptian faience: a blue hippopotamus written over with darker blue.

Maybe the fact that no food is really blue, makes it so special. Only blueberries and concord grapes, dark with blueness and sweetness. Nature saving blue for painters and cloth dyers to discover. Blue houses, blue people and blue robes.

Recently I bought Michael a birthday gift at a local antique/junk store—it's an old *gamelon* instrument, a set of six covered bells, suspended in a wooden grid like a low table. The gongs or bells are meant to be struck with a stick, and form part of the percussive fabric of the Indonesian *gamelon* ensemble. Its music is like night sounds and rivers and bugs and background sounds in a forest, with a rhythm and beat that come from repetition. The wooden support grid is painted blue.

When we found this instrument (and my son and I went back to buy it without Michael's knowing) I marveled that it was painted blue, which is so rare in architecture and furniture in North America, not so in southern and Asian and Caribbean climates. A chalky blue, an old, settled blue. There are cultures where doors are painted blue for protection, and amulets that are blue, and single eyes, meant to keep away evil.

Oxidation. Transformation. The mythic, in art and life. Sky. Sea. The unknown. Blue represents bigness, beauty, a cooling of the passionate heart into peace. Blue roses for my grave, I used to yell, as my turn to jump on the hurtling go-cart going down the high school hill. But now I don't need them.

Kasmi Ink

In making black ink, instead of ferrous actual rust can be substituted, old nails or bits of iron sitting in water. I put water, rusted nails, and molasses in a large ceramic pot in my studio, in proscribed proportions, cover it, and wait to see what happens.

It's been almost two weeks and I can smell the *kasmi*. A dark, molasses-y smell. Every day I lift the heavy earthenware lid, and stir the liquid-and-nails with a wooden chopstick. Then I am gone for 5 days, come back, lift the lid and find a vertical cloud of foam pushing up from the center of the pot, which dissipates as I stir. Fermenting. The water has partly evaporated, leaving a dark line one inch above the level of the liquid. Looking at the recipe for ink I see I will need ground tamarind seed soon, by the end of the week, to mix with the liquid in the right proportion and boil for 30 minutes. This should result in a black ink paste. This morning I decide to go to the market to find tamarind, also to try to find bulgur wheat for tabouli— and walk up and down aisle where the beans and spices and wheats are all displayed in round baskets. Beautiful earth-colored textures, surrounded by baskets. I settle on something that looks like bulgur, but needs grinding. It is light beige instead of reddish brown, perhaps another variety. And I buy two packages of sticky tamarind fruit, dark brown, from which the seeds must be extracted, then pan-dried, then ground. There is no such product here as a package of "ground tamarind seed."

After the molasses and rusty iron sit covered in the corner of my studio for 2 weeks, I measure the liquid, pour it into a small pot and mixed in about 20 grams of tamarind seed powder. Then I boil the mixture on the stove for 30 minutes, stirring with an old wooden salad fork. The liquid looks like bubbling fudge. I will strain it when it is cool, and keep it in a new container in the studio where I will try drawing with it in old and new ways.

A RECIPE FOR KASMI
(courtesy *Aranya*)

Water   10 KG.
Ferrous   10 KG.
Molasses   2 ½ Kg.

Let combination sit in a ceramic covered vessel for 15 days. Stir daily with a chopstick. Strain.

Mix 1 liter liquid with 30-40 grams tamarind seed powder. Boil for 30 minutes in a pan dedicated to dyeing. Ink is ready to use.

The Indigo Show

As I paint this morning, dipping into the tumbler of new really very brownish black ink, despite all the iron, I get the idea to add to what I did yesterday—the unfinished pieces, by adding coffee. I rush into the kitchen to see if there is any left—Michael has switched from tea quite suddenly lately—there is—and pour it into a plastic container. An inspired idea. The coffee turns out, in relation to the ink and the water hyacinth paper, to have a dark, metallic tonality. I have used coffee in drawing classes, but never experimented with it myself. The new ink is warmish, and needs a number of layers to achieve a dark. The two complement each other, and remind me of some of the paintings for *Aranya,* where I used more than one vegetable color dye.

This use of organic color—sometimes a substance right at hand, an everyday material, —or the more complex dyes and inks, which need working up—has a good feel to it. I'm still painting with oils (soon to stop that altogether), but at times the colors begin to feel pushed, unreal, in relation to what I'm using on paper now.

One drawing—this one all hand made dye—has a shaft of light striking its center. The field of the drawing is made up of innumerable marks, which overlap each other, forming a moving stream. Another has very large loose brush marks, in brown/black ink, superimposed with the cooler coffee. The marks hover over each other, not coinciding, thus creating space, and movement. This could be titled *Rain,* like the small abstract painting, a full palette of blues, that I showed at the Fuller Museum's Abstract Show, a Triennial of Boston artists' works. The difficulty I had been having getting back into my work after cancer, and understanding the relationship between the marks in the drawings, so simple, so confident and alive, and the world of my paintings, seems to be dissolving. These marks of organic color link the drawings, which graphically stand by themselves, and the paintings, which take on so much more color, field, space, abstraction.

My thoughts earlier about color and food, and healing, seem to be coming to some conclusion too. When I first came to Bangladesh, despite the fact

I really love all Indian food and enjoy experiments in cuisine, I felt at a loss to reconstruct the nearly vegetarian, fresh diet we eat in the states. The short winter months are the only ones that produce a wide range of vegetables here: broccoli, cauliflower, red cabbage, beets. The rest of the year we live on tomatoes, green beans, cucumbers, carrots and spinach. No imports. But over the last few months I have found bean sprouts, tofu, a dark grainy bread, great yogurt, and a host of fresh spices: basil, coriander, parsley. I sent away for a wonderful, freestanding wok, so food can be stir-fried fresh. And we are planning a garden. Color, texture, spice and story have become part of my life here. The ray of light in my drawing penetrates the dark around it.

I go over to *Aranya* to talk about posters and the upcoming show. Ruby and Moni are in the office. I notice a hand drawn poster on the outer door, with a block print design in indigo paste along the outer borders. I discover that I am to letter 24 of these posters, which will be a challenge as I am not a neat printer. As I sit talking with them someone comes in with traditional Hindu sweets, small hard balls made of grated coconut and molasses, uncooked, which we sample. From the windows of the office we can see Mr. Hossain printing up more poster-borders in dark indigo blue.

I take down the information for the poster, wondering how easy this is really going to be to do:

Aranya & M.C.C.

Present

"Neel Utshob"

Indigo Exhibition

To celebrate the revival of Bengal Indigo

From Fri. 23rd November to Sat. 8th of December

10 am–8 pm

at *Aranya,* 60 Kemal Ataturk Avenue, *Banani*

We talk about this and that, and then I ask if I can peruse the books on Ruby's bookshelf, behind a locked glass panel. So I stand looking at old and new dyeing manuals, as people come and go in the office . Before I leave I wander through the store again, the products always different, always changing, and buy a large indigo tablecloth and napkins for a friend's wedding. Then I go back to my house with borrowed books about indigo, to do some more work.

Another day I come back to work on the long banner that is to hang outside the store for the week of the show, as I have been asked. One day I draw out the shapes of the letters in pencil on the cotton cloth, large: Indigo Exhibition. And now I fill them in with wax: I dip the brush in the pan of wax heating on a burner, then fill in a space in a letter, then go over it again . . . Soon it feels long and tedious. I have utmost respect for the men who are doing this kind of work all the time, although theirs is stamping, not filling in. In the beginning the wax is coming out white, which means the indigo bath will come through—it's too light a coating. With the help of the two technicians I get it: go slower, and keep the brush in the wax longer, just a fraction of time more so the old wax melts as the new wax is picked up by the brush. The letters come out beautiful. The banner will be stamped with a wax pattern around the edges, and dyed in an indigo vat, coming out a deep, pure blue.

The show opens, the shop miraculously ready after apparent chaos the day before. In the courtyard a *shamayana,* or tent is set up, and underneath it on two tables an indigo printing demonstration is going on. The steps of indigo harvest and extraction are shown in photographs along the wall, and there are packets of seed and dried indigo powder on a table. Rubi and Moni are outside, both in beautifully stamped and woven blue saris. The day before the show opened, we artists arrived to put up our work. Hardly an ideal place for showing art, we have the inner entrance room and the walls of the staircase to arrange the work. To make things more difficult, Ruby invited many artists at Dhaka University Fine Arts Department to do indigo paintings at the last minute, and they are there in a pile, unframed. Several of these include the most famous artists in Bangladesh, so we have been supplanted in some ways and there's an awkward feeling as we vie for space for our work.

After much haggling, we work out a situation where the four of us will have work directly in front of and to the right of the door as you go into the store, and in stacks by the side of the window. Later some of the work will be arranged on the staircase. Rokeya's work is soft, with Chagall-like drawing of primitive figures. Alka weaves marks of colored dye in horizon-less landscapes. My work includes two blue stroke paintings, powerful swaths of indigo, several layered paintings which appear dense and nature-like, the experimental dot paintings on silk, and the dense blue *Delta*. Arham's scroll paintings on cotton hang in the upper hallway, and his diverse drawings in indigo. I go up to the upper salesrooms and my breath is taken away by the flood of blue: Ruby and Moni have been setting aside indigo products for awhile and everyone has been working last minute to produce enough to show and sell. Several other craft stores are also represented, as well as "Source," selling hand made books and blue wrapping papers. On the days that I visit the show, there is a strong flow of people moving through, and a sense of excitement about this rare, wild, blue color.

Black Beans and Bangladeshi Bay

Making black beans one cold October afternoon, I pull out whole bay leaves from a plastic bag in the spice drawer: long, slender bay leaves unlike the usual short, stout ones you can buy here in the U.S. I tuck two leaves into the pot of simmering beans—the water is purple and black, from soaking—and the subtle fragrance of bay begins to drift through the kitchen. I am back in Dhaka, in the long white kitchen in Bangladesh, where a huge bag of bay leaves are drying. And outside, where they have been collected, the big generous, overarching bay tree, that created a pool of shadow in the corner of the yard and lent a delicate scent and a sense of the oldness of gardens and plantings to our overdeveloped corner of Dhaka in Bangladesh.

The dried leaves, turned orange in the water, make small canoes as they curl. Several dark beans rest in each.

We have heard from people who took over our home in Bangladesh that Paul, who did odd jobs for us, has been in distress lately. His oldest daughter, of three, has eloped with the boy next door and gone to live up North. Paul had been paying for her to go to special boarding school, and he had hopes for her future as an educated Bangladeshi woman. A middle class person in Dhaka, Paul saved carefully for his family's needs and we helped him too, beyond his salary, for his children's education. Now I imagine him, his tall dark frame leaning over the kitchen counter or pulling off bay leaves in the yard, thinking about the future, and his daughter.

And the unpredictability of life in Bangladesh, or anywhere.

Epilogue

Nature has laid the ingredients of dye all around us in the countryside. Do we notice the gift?

In early dyeing tannic materials derived from seeds, bark and stalks of plants, creating brown and cream colors; leaves and flowers such as henna, safflower, madder created orange, yellow, red. Black was achieved by mixing tannins with natural iron—on skin, first, then eventually fabric.[1]

Synthetic dyes were originally aniline dyes, made from coal tar. Prussian blue and chrome yellow replaced the natural dyes indigo and madder (alizarin). Synthetic dyes involved an easier process but were less stable and their hue was more metallic and sharp, and less subtle.[2]

Tannin, the healthy compound found in black tea, was once used to color and strengthen leather.[3] On Ossabaw Island, we walked sandy causeways to their end, passing pools and streams of water by the side of the paths. These pools were a deep rich brown, the color of tea. Even then I knew I was observing one of Nature's artistic products, the tannin in the bark and sticks of fallen branches all around, causing coloration of the water. I made the connection between tea and the color of the water in the natural pools, between color, nature, and "curing."

Many times during the workshop on natural color I felt impatient, weighing the tiny heads of flowers and dried bark, waiting for a mixture to boil, or cloth that was hanging in the sun to dry; listening to the instructions carefully given to us again and again. Working with natural color can be a slow process, analogous to the process of healing. As we watched, a piece of cotton or silk, or the hand made paper I favor for painting, slowly turned into another color. Lightening, brightening, clearing, clarifying, through color we were trying to understand nature's processes and products, to get to the source of things.

Postscript

From around 1985 to 1995 I chose to paint within a "limited palette," neutral color within a narrow range, especially exploring black and white. Within the range I discovered an infinity of color. This was my intention, to limit my range in order to understand color more; to find more color where there was supposedly less. I had been painting in a high value, high intensity mode for years, emulating Matisse, the Fauves, the Nabis. In the 80's I began to have a nagging feeling that I needed to learn more about color itself and its interaction, not just the interaction of brilliant colors, whose hot, bright natures vibrated intensely together side by side, but subtle relationships and mixtures such as one would see in nature. Fine-tuned differences, colors that hovered within grays. I was seeking a new sense of color, in the dark.

The work that came out of this change, after a year of painting on scraps of canvas to try to discover what I wanted to do, and experimentation on these with new tools and additives had much of the raw, wild side of nature that I had encountered on Ossabaw, during that sojourn in wildness, painting watercolors. In a way, I had found my way into painting those experiences I'd had on the island, the quality of those encounters, in oil but on a larger scale. What I couldn't do right after the island experience (I'd gone back to still life and huge fans, which was as close as I could seem to get to the motion of palm trees) I was now, through a decision about color, immersed in.

I had, since the island, taken photographs of nature. Mostly close-ups, not to use literally as visual material for painting, but to remind me of where I was going. Then, in my factory studio in Somerville where these new dark paintings originated, I put up many photographs on the wall of forest and field texture, willing the energy and depth of nature into my urban setting.

This period of reduced color lasted for about ten years. Like Picasso's blue period, it foreshadowed many difficult life events and revelations, including a miscarriage of a much hoped for birth. It was like a coming to terms, a sweeping away of illusions and bright festivities to what was really there, was part of life.

Now painting with the natural dyes, on larger paper for an upcoming show, I am brought back to the monochrome work of that period, the technique of building up from washes a kind of tapestry of texture and energy that speaks of life in nature; passion; change. Having the door to painting large oils close eventually led me to the making of inks; working on paper; making lighter, less dense paintings with gentler materials. Yet I was still seeking the internal, organic form and energy of growth.

It's just frost time, in the first year we're back from Asia. The garden has provided basil, mint, thyme, parsley, tomatoes, collard greens, squash and peppers—all this despite a very late planting time, our late arrival back in New England. Tall, bright marigolds planted on the edge of the garden can be seen as a pool of yellow light from the back windows. They almost overshadow the vegetables, they are so tall.

As I bring in the geraniums from the hanging pots I pick the dead flowers, and, curious, drop them in a large jar of water. They formed a warm brown liquid, which I use as a base in some of my brush paintings. I re-pot the geraniums for winter, inside; pull all the basil, which becomes pesto; and hang bunches of mint to dry in the barn, below my studio. This is small potatoes gardening, but each aspect of it gives me ideas, like the darkening and chilling of winter, where light and warmth goes within.

I welcome this chance to go within, like the Persephone myth that mirrors the process of winter and spring, dark and light: the daughter of Demeter, the earth, is forced to spend half the year in the underworld, through a bargain made with Hades.

We are filling our orange-yellow kitchen, with its large sitting area, the place we all spend most of our time together, with the geranium pots; narcissus bulbs, coming up in water; and a new orange hibiscus. The yellow flowers are still blooming in the garden.

APPENDIX

The Secrets of Thai Color

Thailand tantalizes with its colorful lifestyle, its radiant architecture and wild street life, but after a year and a half the progress on researching natural color is slow. One book, two newspaper articles, brochures from Chiang Mai. After several years back in the U.S., I had returned to Asia, settling into an apartment in Bangkok on a dead-end street for quietness and trees, wondering what this new sojourn would bring, how it would be different from living in Dhaka. Since I had not been able to bring natural dye materials with me to make inks, due to a small baggage allotment and fear of seizure by customs, I expected I would find common color-producing materials right away.

In search of natural color information, I scan the web; I ask local markets; I ask everyone I meet. In handmade and hand-dyed clothing stores filled with a multiplicity of blues, I ask where they sought indigo. "We have sources," I am mysteriously told. Or, "in the Northeast, in impoverished *Isan*," and "don't ask—no one will share." I ask whether I can buy some indigo cake or powder? "No—we don't deal with the suppliers directly." I hear of someone cultivating and using indigo in Chiang Mai, a friend of my Bangladeshi friend Ruby's, and e-mail her. I try again. I call. I call again. No response.

Finally, I search the web for textile suppliers in the U.S. who will ship overseas, and have *cutch* (acacia), brazilwood, and the necessary Spectralite sent from Asheville, North Carolina. Anil and Jyoti, friends in Dhaka, bring some indigo for me, (it is easy to buy in Dhaka), the next time they come to Bangkok. I find a local supplier of Gum Arabic who gives me free samples, and buy the needed washing powder at a grocery store. A mortar and pestle is easy to find, and eventually I locate a tiny scale in Chinatown that will measure small amounts. I meanwhile paint with turmeric, easy to find in an Indian store, and tea and coffee from my own kitchen.

Why is it so hard to find the natural dye ingredients here? A sign at the Siam Society, a large and beautiful cultural complex on *Asoke* (Avenue), just off *Sukhumvit* reads,

Lanna women often kept their dye formulas secret. Dyes were extracted from wood, bark, root, leaves, seeds, fruit and flowers from the village and forest environment, with the more exotic reds and pinks obtained from the resin of the *krang* or *lac* insect. To fix the dyes, mordants made from substances as varied as tamarind juice, cow dung, mud and bark ash.[1]

Reading this, I understand the history of dye making in Thailand seems to have been an oral transmission of dye recipes handed down from generation to generation. As the industry was local, the techniques and secrets are closely guarded.

Just as I was getting very frustrated and ready to give up and order from abroad the materials to make color, I took a three day trip up north with an Australian woman who raised money for remote schools in Thailand near the Burmese border. We drove for hours along twisting, winding roads amidst sheer drops on either side, breathtaking landscapes, soft blue hills overlapping some of the most mountainous regions of Thailand. We ate by lantern light with teachers at their barracks, munched on frog curry and pork rind outside a school, and wandered through *Karen* villages, with houses on stilts. On the way back to Chiang Mai my companion asked if I wanted to visit an interesting museum and industry. She said they make "color"—and before long we drove down a long roadway lined with thick bamboo to the *Pa-Da* Cotton Textile Museum and *Baan Rai Pai Ngarm* industry, where pure cotton cloth was spun and naturally colored with dyes.

On the shore of a slow, winding river, whose orange and green banks uncannily were reflected in the green and orange being spun on the outdoor looms underneath the museum rooms, were cottons of unimaginable subtlety of color, lovingly displayed in cases. Around a courtyard were vat after vat of indigo, of varying strength, stained deep dark blue around the edges. The whole place had a sleepy, past-its-time air. The woman who had started this place, Saeng-da Bunsiddhi, intended to keep the ancient dye-making and weaving going and keep this tradition alive for the future. After her death, her granddaughter, Klaowarat Bunsiddhi, runs it. I picked out two meters of a beautiful cloth with many blues and greens running through it, to bring back with me to Bangkok.

I couldn't help but reflect on the difference between Bangladesh and Thailand. Bangladesh, still rebuilding itself after its bloody independence war of 1972;

struggling to make jobs and grow its economy, where people gladly gave me recipes and shared information about color. Thailand, a more developed country, where the combination of traditional family secrets and recent commercialism keeps the information relatively inaccessible. In Bangladesh there was transparency, a lack of concern about economics and commercialism. Instead I found a delight and pride in sharing what was just a beginning resurgence of the ancient art of making natural color. Their only reticence coming from a deep feeling of sadness, shame and repugnance at the dark history of indigo production in Bengal under the colonial hand of the British East India Company, (see chapters *Indigo* and *The Mirror of Indigo*). In Thailand, so much more peaceful and prosperous, commerce has been successfully in place for years. Bangkok is so westernized that people are suspicious of someone asking questions, perhaps to steal a recipe? or compete in selling indigo products? And Thailand also has a tradition of keeping the dye recipes secret and passing them down from family to family in order to keep them pure, and close.

Yet I have learned two new aspects of color in Thailand, both of which I appreciate for their uniqueness. One is the practice of dyeing Buddhist monks' robes with the bark of the jackfruit tree (*khanun*). Jackfruit is the large knobby fruit I originally saw in Bangladesh and now see again in Thailand. Monks gather the bark off jackfruit trees. The chips of bark are soaked for hours to produce a dye as strong as tea or coffee. White robes are soaked in a hot solution until they reach a bright yellow gold, that over time darkens to the traditional forest color. I saw many variations: orange, neutral brown-green, lighter yellow. I learned from a monk at the World Buddhist Association that every two weeks they wash the robes in the same dye, which has an astringent, soapy texture. It is even said to have antiseptic qualities. They put the robes in huge metal bowls called *galamands* over a large log fire, which creates smoky, thick grey air. They open a metal flap on the bowl to get more oxygen, which purifies the fire and gets rid of the smoke. To the monk the smoke seems to describe our lives and even meditation, how most of us assume we can solve our problems through thinking, which tends to make everything more murky, like living in a room full of smoke, when we can't possibly solve anything this way. What we need to do, he said, is open the flap to let in more oxygen, or attention, or breathing into the body. Bringing awareness to the problem, clears it. Breathing calms the mind.[2]

Making colors from the earth in Thailand also has an element of ritual and magic to it, especially around indigo and *lac* (red) dyeing. With both, the proscriptions around polluting the color pot are numerous. Nearby deaths, visits by family members of someone who is sick, or hopping over a dye pot all affect the color negatively. Dyers need to work alone over a dye pot, or drink some of the *lac* dye to insure a good color. No monks, pregnant women or very old people are allowed to come near the pot, and the dyer is not supposed to speak with anyone when bringing leaves to the pot, nor move the wood of the fire or make colors on a Buddhist holiday. The pot is not only secret, but alive with spirits; and the makers of indigo are believed to become ghosts or spirits when they die. A failing dye pot can be remedied by putting a thorny branch on top of the pot to contain the energy or spirit. Mysteriously, the indigo pot is fed with such substances as red ant water, sugar, whiskey, and lemon, presumably to control the Ph, but perhaps to appease the spirits as well.[3]

These beliefs remind me of my feeling in Georgia, USA, where a myth about indigo also resonated with its history, traditions, and power.

I eventually settled into Bangkok, and having accepted the limitations of the city and the culture's attitudes about sharing color knowledge and traditions, I found more material available. The Textile Society held a weekend workshop at *Ayoraya,* a clothing shop off Petchburi Road, on indigo, and I was able, with a small group of people, to view the indigo pot and Thai dyeing methods close up. Later, visiting Vientiane, Laos, with its feel much closer to Dhaka, I discovered several small natural color clothing shops connected to modest dyeing industries in the countryside producing a range of organic material. A show at 100 Tonson Gallery by Yanawit Kunchaethong, "Botanical Art Exhibition," exhibited prints and paintings made primarily from crushed fruits grown on Yanawit's family land in Petchburi. Small jars of organic substances used in printmaking sat next to the art work, along with sketches and test strips of color experiments. We had a brief conversation at the opening in which I learned of his painstaking methods of extraction and experimentation, which curiously had no connection with dyeing. He seemed to share the same reverence for nature I have, and hopes to share this ecological viewpoint through his work.

During the time I was attempting to discover the natural color of Thailand I settled into my Bangkok apartment studio, making an extensive *Raintree Series* of large marks falling vertically out of the abstracted sky. They were a reflection, perhaps, of the enormous rains that fall in Asia, and the rhythmic beat of nature which I could hear softly as it pulsed underneath the noise of the city. When I first arrived, I dreamed of huge trees and a dark earthy pitch, like bark or earth. In the *Raintree Series,* a dark fertile richness resonates through *kasmi, cutch* and indigo colors, earth notes mingling in a dream of Thailand.

Bibliography

Ash, Beryland and Anthony Dysen. *Introducing Dyeing and Printing,* NY: Watson Guptill Publications, 1970.

Bonta, Marcia. *Dyes From Nature.* Brooklyn Botanical Garden, 1990.

Bremness, Lesley. *Herbs.* London: Dorling Kindersley, 1994.

Buchanan, Rita. *A Dyers Garden.* Colorado: Interweave Press, 1995.

Casselman, Karen Leigh. *The Craft of the Dyer.* NY: Dover Publishing, 1980.

Colonial Dames of America. *Herbs and Herb Lore of Colonial America.* NY: Dover Publications, Inc., 1995.

Dendel, Esther Warner. "Blue Goes For Down—How Indigo Dye Came to Liberia—a folk tale." *Natural Plant Dyeing.* Brooklyn, NY: Brooklyn Botanic Garden, 1973.

Faragher, John Mack, ed. *The American Heritage Encyclopedia of American History.* NY: Henry Holt and Co., 1998.

"Indigo." *World Book Encyclopedia.* Chicago: World Book, 2000.

Ghuzavi, Sayijada R. *Rangeen: Natural Dyes of Bangladesh.* Dhaka: Eli San Printers, 1987.

Jaffrey, Madhur. *World Vegetarian.* NY: Clarkson/Potter Publishers, 1999.

Kahn, Moni. *Natural Dye Training.* Aranya: Sept. 2, 2001.

Kanti, Ellen Ph.D., RN, HNC and Dr. Eugene Zampieron, NDS, AHG. *The Natural Medicine Chest.* NY: M. Evans and Co., Inc., 1999.

Kierstead, Sallie Pease. *Natural Dyes.* Boston: Branden Press, 1950.

Kling, Blair B. *The Blue Mutiny.* Philadelphia: University of Pennsylvania Press, 1966.

Moeyes, Marjo, *Natural Dyeing in Thailand.* White Lotus Press, 1993.

Myers, Diana K. "Dyeing in the Himalayas." *Dyes From Nature.* Brooklyn, NY: Brooklyn Botanical Garden, 1990.

Naha, Arun and Joe Ward, *Asia Study Group Tour.* Dhaka: Nov. 18, 2001.

"Nikomal." (flier) Mohammadpur, Dhaka: MCC.

Novak, James J. *Bangladesh: Reflections on the Water.* Bloomington: Indiana University Press, 1993.

Panjabi, Camellia. *The Great Curries of India.* NY: Simon + Schuster, 1995.

Phillips, Roger and Anthony Dysen. *Introducing Dyeing and Printing.* NY: Watson Guptill Publications, 1970.

Ross, Gary V. "A Blue Future for Mexican Indigo." *Dyes From Nature.* Brooklyn, NY: Brooklyn Botanical Garden, 1990.

Saffer, J. Daniel. "Wild Plant Dyes." *Dyes From Nature.* Brooklyn, NY: Brooklyn Botanical Garden, 1990.

Sandberg, Gosta. *Indigo Textiles: Techniques and History.* London: A + C Black, 1989.

Siddiqui, Lt. Col. Faiz. *The Lost Nilkuthis of Jessore-Kushtia (1795-1895).* Dhaka: Jesmeen Ahmed, 2001.

Tannahill, Reay. *Food in History.* NY: Stein and Day, 1973.

Webster's New World College Dictionary 3rd Edition. USA: MacMillan, 1996.

Wells, Kate. *Fabric Dyeing and Printing.* London: Interweave Press, 1997.

References

INDIGO

1 Nilkomol, MCC (flier), Mohammadapur, Dhaka, p.3.
2 Lt. Col. Faiz Siddiqui, *The Lost Nikuthis of Jessore-Kushtia,* Dhaka: Jesmeen Ahmed,2001, pp. 41–46.
3 John Mack Faragher,ed., *The American Heritage Encyclopedia of American History,* NY: Henry Holt and Co., 1998, p.445.
4 Nilkomol, p.1.
5 *World Book Encyclopedia,* Chicago: World Book, Inc., 2000, p.223.
6 Faragher, p.467.
7 Beryland Ash and Anthony Dysen, *Introducing Dyeing and Printing,* NY: Watson Guptill Publications, 1970, p.113.

BASIL

1 Roger Phillips and Martyn Rix, *Herbs for Cooking,* NY: Random House, 1998, pp.14,16.
2 Lesley Bremness, *Herbs,* London: Dorling Kindersley, 1994, p.260.
3 Rita Buchanan, *A Dyer's Garden,* Colorado: Interweave Press, 1995, pp.84–85.

PAINTING WITH NATURAL DYES

1 Esther Warner Dendel, "Blue Goes for Down—How Indigo Dye Came to Liberia—a folk tale," *Natural Plant Dyeing,* NY: Brooklyn Botanic Garden, 1973, pp. 23–28.
2 Bremness, p.95.

CURRY

1 Madhur Jaffrey, *World Vegetarian,* NY: Clarkson/Potter Publishers, 1999, pp.707–708, 721.
2 Camellia Panjabi, *The Great Curries of India,* NY: Simon + Schuster, 1995, p.32.
3 Reay Tannahill, *Food in History,* NY: Stein and Day, 1973, p.165.

4 Punjabi, p.33.

5 Tannahill, p.165.

6 Ibid, p.241.

7 Ibid, p.312.

8 Panjabi, p. 40.

9 Bremness, p. 162.

MINT

1 Phillips and Rix, pp.18, 25.

2 Bremness, p. 190.

PEPPER

1 Tannahill, p. 240.

2 Ibid, p. 101.

3 Bremness, p. 283.

4 Tannahill, p. 55-56.

5 Panjabi, p. 36.

6 Tannahill, p. 237.

7 Ibid, p. 240.

8 Idem.

GREEN TEA

1 Dr. Eugene R. Zampieron and Ellen Kanti, *The Natural Medicine Chest,* NY: M. Evans and Co., Inc., 1999, p.30.

2 Bremness, p. 96.

3 Idem.

4 Zampieron and Kanti, p. 96.

5 Ibid, pp. 96-97.

6 Tannahill, p. 306.

7 Zampieron and Kanti, pp. 97-98.

8 Tannahill, p. 307.

9 Ibid, pp. 309, 344.

10 James J. Novak, *Bangladesh, Reflections on the Water,* Bloomington, University of Indiana Press, 1993, pp. 9-10.

THE WORKSHOP

1 Moni Kahn, Natural Dye Training, *Aranya,* Sept. 2, 2001.
2 Sayijada R. Ghuzavi, *Rangeen: Natural Dyes of Blangladesh,* Dhaka: Eli San Printers, 1987, pp. 24,41.
3 Ibid, pp. 24, 41.
4 *Webster's New World College Dictionary 3rd Edition,* USA: MacMillan, 1996, p. 423.
5 Zampieron and Kanti, p. 143.
6 Idem.
7 Karen Leigh Casselman, *The Craft of the Dyer,* NY: Dover Publications, 1980, pp. 24-25, 27.
8 Ibid, p. 35.
9 Ghuzavi, p. 41.

THE NINE STEPS OF DYEING

1 Bremness, p. 37.
2 Sallie Pease Kierstead, *Natural Dyes,* Boston: Branden Press, 1950, p. 79.
3 J. Daniel Saffer, "Wild Plant Dyes," *Dyes From Nature.* Brooklyn Botanical Garden, 1990, p. 66.
4 Idem.
5 Diana K. Myers, "Dyeing in the Himalayas," *Dyes From Nature,* Brooklyn Botanical Garden, 1990, p.11.

PAINTING NOTES

1 Gosta Sandberg, *Indigo Textiles,* London: A+C Black, 1989, p.111.

PLANTING BLUE GOLD

1 Marcia Bonta, *Dyes From Nature,* Brooklyn Botanical Garden, 1990, pp. 33-34.
2 Myers, p. 13.
3 Colonial Dames of America, *Herbs and Herb Lore of Colonial America,* NY: Dover Publications, Inc., 1995, p. 8.
4 Gary V. Ross, "A Blue Future for Mexican Indigo," *The Dyes From Nature,* Brooklyn Botanical Garden, 1990, p. 37.
5 Sandberg, pp. 14, 19.
6 Novak, p.43.

7 Arun Naha, Sr. Development Officer, Job Creation Program and Joe Ward, Acting Program Leader, MCC, Dhaka: Asia Study Group Tour, Nov. 18, 2001.

8 Idem.

9 Myers, p. 13.

10 Joe Ward, Nov. 18, 2001.

11 Bremness, p. 109.

12 Idem.

MIRROR OF INDIGO

1 Blair B. Kling, *The Blue Mutiny*, Philadelphia: University of Pennsylvania Press, 1966, p.60.

2 Ibid, pp.200–201.

3 Ibid, p. 118.

4 Novak, p. 81.

5 Kling, p. 198.

6 Ibid, p. 198–199.

7 Ibid, p. 201

8 Ibid, p. 203.

9 Ibid, pp.203–204.

10 Ibid, p. 206.

11 Ibid, p. 207.

12 Kling pp.30–31.

13 Ibid pp.18–19, 33, 39, 53–54.

14 Ibid, p.19.

15 Ibid, p. 66.

16 Ibid, pp.61, 67.

17 Ibid, p. 61, 68, 110–111.

18 Ibid, p. 65, 72.

19 Novak, p. 81.

20 Kling, p. 174–78.

21 Ibid, p. 210.

22 Ibid, p. 220.

23 Ibid, p. 193.

24 Novak, p.81.

BLUE

1 Ash and Dysen, p.113.
2 Jess Stein, ed. Random House Dictionary, NY: Random House, 1967, p.161.
3 Sandberg, p.35.

EPILOGUE

1 Kate Wells, *Fabric Dyeing and Printing,* London: Interweave Press, 1997, pp. 8-9.
2 Ibid, p. 12.
3 Zampieron and Kanti, p. 30.

APPENDIX

1 The Siam Society, 131 Soi 21, (Asoke), Sukhumvit Road, Bangkok 10110, Thailand.
2 The World Buddhist Fellowship of Buddhists Headquarters, 616 in Benjasiri Park, Sukhumvit 24 Rd., Klongtoey, 10110, Bangkok.
3 Marjo Moeyes, *Natural Dyeing in Thailand,* Bangkok: White Lotus, 1993. pp. 27, 40-41; 51, 57, 63.

Glossary of Terms

Aranya "forest" in *Bangla;* name of the natural dye store

Banani a suburb of Dhaka where the natural dye store is located

Bangla language spoken in Bangladesh; center of the "Language Movement" in 1971 during the War of Liberation

batik cloth printed with wax, then dyed, then ironed

Bengal the region including eastern India and Bangladesh; before Partition the most celebrated part of India, known for its artists, poets and musicians

cha dokar tea shop

gamelon Southeast Asian instrument ensemble, including percussion and string instuments

Garo tribe living in the hill tracts of Bangladesh

Gulshan suburb of Dhaka where the author lived

Kulna southern central city in Bangladesh

kafta man's loose long shirt

lalsag dark green spinach with bright red veins

lungi long tube of cloth traditionally worn over the legs by men in Bangladesh, tucked in at waist

nilkuthis "blue houses" or indigo factories once used for indigo production in Bengal

orna scarf thrown over the shoulders, traditional dress for women in Bangladesh

Rangamati hill tract area in northeastern Bangladesh

shamayana large tent erected for celebrations in Bangladesh

samosa potatoes, peas and other vegetables covered with pastry and fried, traditional Bangladeshi and Indian snack

shalwar kameez outfit worn by women in Bangladesh: baggy pants, long dress with slits at the sides, *orna* scarf thrown over shoulders

taka unit of money in Bangladesh; 58 *taka* equivalent to 1 US dollar

Sarah Sutro is a painter and writer whose work has been shown and collected extensively in the U.S. and internationally. She has been a fellow at the MacDowell Colony, Blue Mountain Center and a visiting artist/writer the American Academy of Rome. She is a recipient of a Pollock Krasner Grant and was a finalist for the Robert Frost Poetry Award. As well as *Colors: Passages through Art, Asia and Nature,* her poems are published in the Thailand anthology *Bangkok Blondes;* she is the co-editor of *Buddhist Chanting;* and she collaborated on the photographic essay *Through our Eyes* published by the National Museum of Thailand. An affiliate professor with Union Institute & University, she has taught in over 10 colleges and universities. She has lived and travelled for much of the last decade in Asia, and currently lives in Western MA (USA).

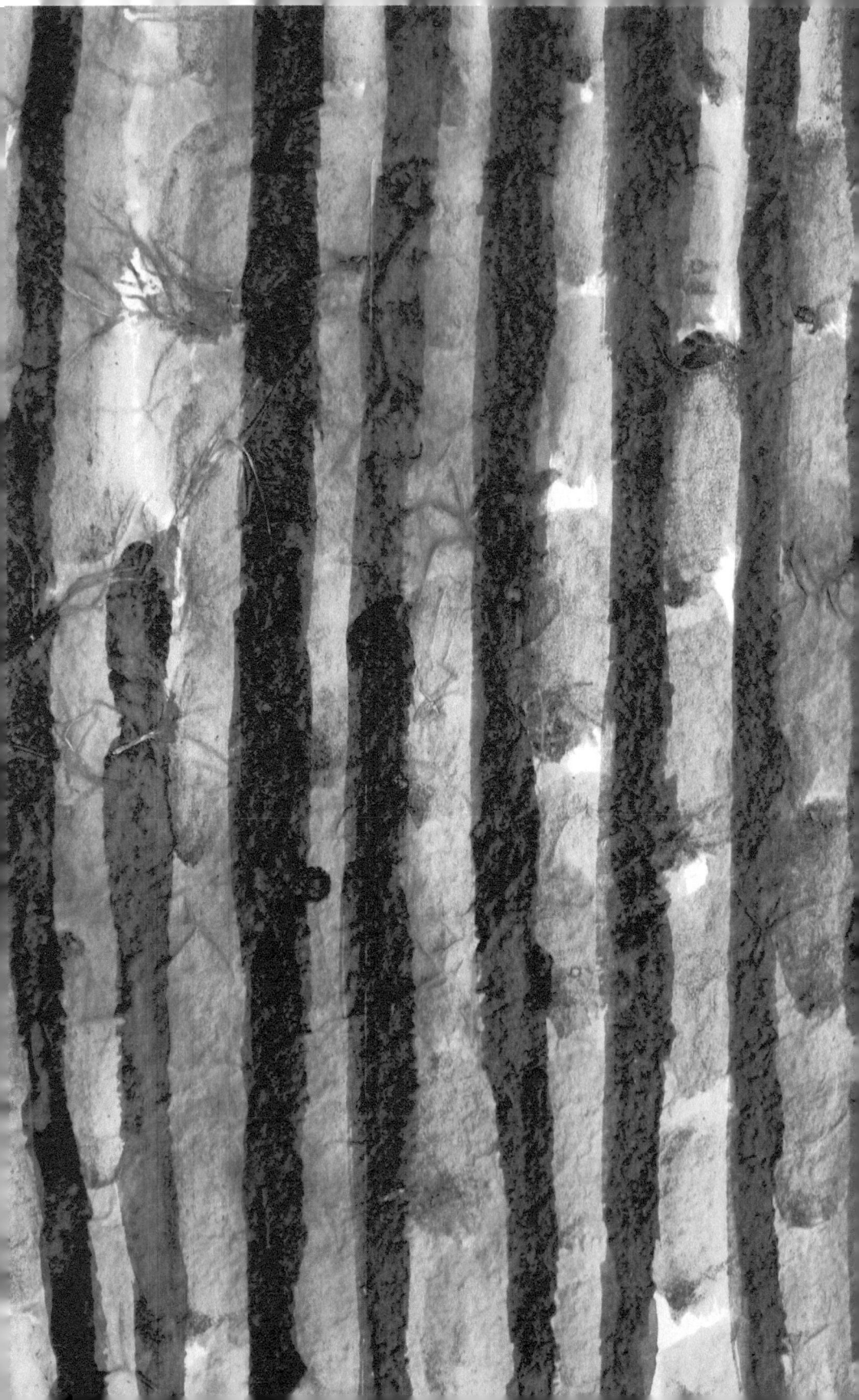

www.ingramcontent.com/pod-product-compliance
Lightning Source LLC
Chambersburg PA
CBHW022002170526
45157CB00003B/1106